Denis Gonta

Generation of entangled states in the framework of cavity-QED

Denis Gonta

Generation of entangled states in the framework of cavity-QED

Schemes for generation of multipartite entangled states with atomic chains in the framework of cavity quantum electrodynamics

Südwestdeutscher Verlag für Hochschulschriften

Impressum/Imprint (nur für Deutschland/ only for Germany)
Bibliografische Information der Deutschen Nationalbibliothek: Die Deutsche Nationalbibliothek verzeichnet diese Publikation in der Deutschen Nationalbibliografie; detaillierte bibliografische Daten sind im Internet über http://dnb.d-nb.de abrufbar.

Alle in diesem Buch genannten Marken und Produktnamen unterliegen warenzeichen-, marken- oder patentrechtlichem Schutz bzw. sind Warenzeichen oder eingetragene Warenzeichen der jeweiligen Inhaber. Die Wiedergabe von Marken, Produktnamen, Gebrauchsnamen, Handelsnamen, Warenbezeichnungen u.s.w. in diesem Werk berechtigt auch ohne besondere Kennzeichnung nicht zu der Annahme, dass solche Namen im Sinne der Warenzeichen- und Markenschutzgesetzgebung als frei zu betrachten wären und daher von jedermann benutzt werden dürften.

Verlag: Südwestdeutscher Verlag für Hochschulschriften Aktiengesellschaft & Co. KG
Dudweiler Landstr. 99, 66123 Saarbrücken, Deutschland
Telefon +49 681 37 20 271-1, Telefax +49 681 37 20 271-0
Email: info@svh-verlag.de
Zugl.: Heidelberg, University, Diss., 2010

Herstellung in Deutschland:
Schaltungsdienst Lange o.H.G., Berlin
Books on Demand GmbH, Norderstedt
Reha GmbH, Saarbrücken
Amazon Distribution GmbH, Leipzig
ISBN: 978-3-8381-1955-7

Imprint (only for USA, GB)
Bibliographic information published by the Deutsche Nationalbibliothek: The Deutsche Nationalbibliothek lists this publication in the Deutsche Nationalbibliografie; detailed bibliographic data are available in the Internet at http://dnb.d-nb.de.

Any brand names and product names mentioned in this book are subject to trademark, brand or patent protection and are trademarks or registered trademarks of their respective holders. The use of brand names, product names, common names, trade names, product descriptions etc. even without a particular marking in this works is in no way to be construed to mean that such names may be regarded as unrestricted in respect of trademark and brand protection legislation and could thus be used by anyone.

Publisher: Südwestdeutscher Verlag für Hochschulschriften Aktiengesellschaft & Co. KG
Dudweiler Landstr. 99, 66123 Saarbrücken, Germany
Phone +49 681 37 20 271-1, Fax +49 681 37 20 271-0
Email: info@svh-verlag.de

Printed in the U.S.A.
Printed in the U.K. by (see last page)
ISBN: 978-3-8381-1955-7

Copyright © 2010 by the author and Südwestdeutscher Verlag für Hochschulschriften Aktiengesellschaft & Co. KG and licensors
All rights reserved. Saarbrücken 2010

Preface

Cavity quantum electrodynamics (cavity-QED) is an established research field that studies electromagnetic fields in confined spaces and radiative properties of atoms in such fields. Experimentally, the simplest example of such system is a single atom interacting with modes of a high-finesse resonator. Theoretically, such system bears an excellent framework for quantum information processing in which atoms and light are interpreted as bits of quantum information and their mutual interaction provides a controllable entanglement mechanism. In the recent years, a remarkable progress has been achieved with regard to fabrication of high-finesse resonators and coupling to them various atomic systems or even macroscopical objects. These achievements marked a new chapter in the physics of coherent light-matter interactions and triggered novel experimental developments in the area of quantum optics.

This manuscript is based on the dissertation submitted to the Combined Faculties of Natural Sciences and Mathematics of the Ruperto-Carola-University of Heidelberg for the degree of Doctor of Natural Sciences. In this manuscript, I present several practical schemes for generation of multi-partite entangled states for chains of atoms which pass through one or more high-finesse resonators. First, I propose two schemes for generation of one- and two-dimensional cluster states of arbitrary size. These schemes are based on the resonant interaction of a chain of Rydberg atoms with one or more microwave cavities. Secondly, I propose a scheme for generation of multipartite W states. This scheme is based on the off-resonant interaction of a chain of three-level atoms with an optical cavity and a laser beam. I describe in details all the individual steps which are required to realize the proposed schemes and, moreover, I discuss several techniques to reveal the non-classical correlations associated with generated small-sized entangled states. All presented schemes can be adapted to the present day or near-future developments in cavity-QED.

Contents

Introduction 1

I Theory of the atom-light interaction   5

1 Interaction of a two-level atom with a single-mode cavity field   7
   1.1 Quantized light field in a planar cavity 8
       1.1.1 Transverse cavity field components 12
   1.2 Atom coupled to a cavity 14
       1.2.1 Resonant atom-cavity interaction 17
       1.2.2 Effective atom-cavity interaction time 20
       1.2.3 Damping of Rabi oscillations 23
   1.3 Summary 25

2 Interaction of three-level Λ-type atoms with cavity and laser fields   27
   2.1 Semiclassical atom-field interaction 28
   2.2 Generation of multipartite entangled states 31
   2.3 Combined atom-cavity-laser interaction 35
       2.3.1 Effective single-mode Hamiltonian 36
       2.3.2 Combined Hamiltonian and the Schrödinger equation 38
   2.4 Far off-resonant interaction regime 39
       2.4.1 Effective Hamiltonian and the asymptotic coupling 40
   2.5 Summary 42

II Cavity-QED experimental setups   45

3 Microwave cavity setup   47
   3.1 Microwave cavity 48
   3.2 Circular Rydberg atoms as qubits 49
       3.2.1 Ramsey plates 50

3.3 Summary 53

4 Optical cavity setup   55
4.1 Optical cavity 56
4.2 Neutral atoms as qubits 57
    4.2.1 Transportation of atoms 58
4.3 Summary 59

III Multipartite entangled states for chains of atoms   61

5 Generation of entangled states with a microwave cavity   63
5.1 Entangled states with a single-mode cavity 65
    5.1.1 W states 66
    5.1.2 GHZ states 67
    5.1.3 Linear cluster state 70
5.2 Proving the entanglement generation 73
    5.2.1 Entanglement measure for an atomic Bell state 76
    5.2.2 Three-partite entangled GHZ and W states 78
    5.2.3 Entanglement measure for a cavity Bell state 81
    5.2.4 Four-partite entangled GHZ 82
5.3 Two-dimensional cluster states 84
    5.3.1 $2 \times N$ cluster state 85
    5.3.2 $3 \times N$ and arbitrary two-dimensional cluster states 89
5.4 Remarks on the implementation of schemes 92
5.5 Summary 93

6 Generation of entangled states with an optical cavity   95
6.1 W states for atoms conveyed through a cavity 96
    6.1.1 $N = 2$ partite state 97
    6.1.2 $N = 3$ partite state 99
    6.1.3 $N = 4$ partite state 101
    6.1.4 $N \geq 5$ partite states 102
6.2 Remarks on the implementation of schemes 105
6.3 Summary 106

Outlook and Acknowledgements   109

Introduction

The pioneering proposal to perform a computation according to the laws of quantum mechanics was initially suggested by R. Feynman in the 1980s. He realized that it gets extremely difficult and infeasible to simulate quantum systems by using conventional computers since the required processing power increases exponentially with the size of system. He made, therefore, one radical proposal to simulate the quantum mechanics by using the quantum hardware. Later on, D. Deutsch outlined the basic principles of quantum computation in Ref. [1] in which the central idea was to encode the bits of information as quantum states. He anticipated that quantum computation might outperform the classical computation if one exploits the ability of a quantum mechanical system to exist in a superposition of two distinguishable states.

At that time, however, quantum computers were considered not more than an academic curiosity. A key breakthrough was made in 1994 when P. Shor showed in Ref. [2] that a quantum computer can factorize a large number into primes in a polynomial time rather than exponential time. Since this factorization has been used as a basic ingredient in various cryptographic protocols, the quantum computing has attracted an enormous attention and interest. After Shor's breakthrough, moreover, further evidence for the outperforming power of quantum computers came in 1996 when L. Grover showed in Ref. [3] that the problem of searching through an unstructured database could also be done much faster on a quantum computer. Owning to these and others fascinating concepts, the field of quantum computation and quantum information has been growing at amazing pace and has become an established branch of research in physics with connections to mathematics and computer science [4].

In all the mentioned applications, moreover, the correlated superpositions of multipartite states – the (so-called) entangled states, which involves two or more interacting quantum sub-systems, play an essential role. Despite the fact that the notion of entanglement have been known since the early days of quantum mechanics (see Refs. [5, 6]), it took nearly thirty years until A. Aspect revealed experimentally the nonlocal character of simplest two-partite entangled states [7]. Although a variety of more refined experiments have been carried during the recent years, there is still an ongoing debate about the question if our world is actually nonlocal at the microscopic level. We can conclude,

INTRODUCTION

therefore, that despite of its puzzling and counterintuitive implications, the quantum entanglement has been found essential not only in studying the non-classical behavior of composite quantum systems but also as one vital resource in quantum information processing.

During the last decades, quantum optics turned to be one of the most rapidly developing areas of modern physics in which the concepts of quantum information are manifested in the most spectacular way. The general trend of this progress can be characterized by increasing of the precision to manipulate single quantum systems and, moreover, by providing controllable entanglement mechanisms for several such quantum systems. While the physical realization of basic quantum gates and algorithms in the framework of quantum optics has been achieved [8, 53], the scaling of these schemes to larger systems remains still a great challenge. The major difficulty originates to the fragility of quantum systems caused by the interaction with the environment and which leads to the decoherence [9]. In atomic systems, moreover, the main source of decoherence is spontaneous emission of the excited atomic state caused by its coupling to the free-space electromagnetic background. One of the techniques developed to control the spontaneous emission leads us into the area of cavity quantum electrodynamics (cavity QED) which studies electromagnetic fields in confined spaces and radiative properties of atoms in such fields [10].

Cavity-QED emerged in the 1970s with experimental studies of how radiative properties of atoms are modified when they radiate close to boundaries, i.e., when the atom is placed inside a closed cavity [11]. The dynamics of coupled atom-field system, however, remained unexplored since the photons were lost much faster than the characteristic interaction times. With the better resonators which were developed later on, a new epoch in cavity-QED has been marked. Namely, the coupling of an atom to the cavity mode has become a dominant effect in the atom-field evolution [12]. Radiative properties of atoms in this (so-called) strong coupling regime significantly differ from what was observed before. Spontaneous emission, for instance, is replaced by periodic Rabi oscillation and becomes thus a reversible process [34].

In the recent years, furthermore, a remarkable progress has been achieved with regard to fabrication of high-finesse resonators and coupling to them various atomic (ionic) systems or even macroscopical objects. These achievements marked a new chapter in the physics of coherent light-matter interactions and triggered novel experimental developments in the area of quantum optics. On the other hand, cavity-QED bears an excellent framework for quantum information processing in which atoms and cavity photon field are interpreted as bits of quantum information (qubits) and their mutual interaction provides an exceptional entanglement mechanism [38, 39, 40, 41, 42]. Owning to this entanglement mechanism, therefore, in this manuscript we present several

INTRODUCTION

practical schemes for generation of multipartite entangled states for chains of atoms which pass through one or more high-finesse resonators.

In the first step, we propose schemes for generation of one- and two-dimensional cluster states which represent a novel type of multi-partite entangled state introduced by H. J. Briegel and R. Raussendorf in Ref. [48]. Apart from the fundamental interest in these states [50] and their use in quantum communication protocols [51], the cluster states are the key ingredient for one-way quantum computations [52]. In the recent years, the generation of cluster states has attracted much attention and has become a research topic by itself. Using a linear-optical set-up, for example, a proof-of-principle implementation of a four-qubit one-dimensional (linear) cluster state has first been reported [53] and utilized in order to demonstrate basic operations for the one-way quantum computing [54]. In the framework of cavity-QED, moreover, different schemes have been suggested to generate linear cluster states [47, 55, 56, 57, 58]. In contrast to the linear (one-dimensional) cluster states, however, the two-dimensional cluster states enables one to perform also the quantum gates which act on two or more qubits simultaneously [52] and, therefore, it may result in a viable alternative to the conventional (circuit) computations. Up to the present, nevertheless, only a minor progress has been done in Ref. [59] in order to generate small-sized two-dimensional cluster states in the framework of cavity-QED.

In the third part of this manuscript, we describe our scheme to generate the linear $(1 \times N)$ cluster state, and right afterwards, two schemes to generate the two-dimensional $2 \times N$ and $3 \times N$ cluster states. These schemes work in a completely deterministic way and are based on the resonant interaction of a chain of Rydberg atoms with one or more microwave high-finesse cavities which support two independent modes of photon field. While only one of these cavities is required for the generation of linear and $2 \times N$ cluster states, two (and more) cavities are needed to generate cluster states of larger size [61]. For each scheme, we describe the individual steps in the interaction of each atom with one of cavity modes. We also make use of a graphical language in order to display all these steps in terms of quantum circuits and temporal sequences. In addition, we show how the last proposed scheme can be extended to generate two-dimensional cluster states of arbitrary size once a sufficiently large chain of atoms and an array of cavities are provided. We briefly discuss the implementation of one-way quantum computations by considering the setup similar to those utilized in the Laboratoire Kastler Brossel (ENS) in Paris [34], and we conclude that our schemes are well suited for the present-day developments in cavity-QED.

In the above schemes, the atoms interact resonantly with one of cavity modes while passing sequentially through an array of cavities such that only one atom is coupled to a mode at a given time. One totally different regime of interaction can be realized if

INTRODUCTION

a chain of two or more atoms is placed inside the cavity such that all the atoms are simultaneously coupled to a slightly detuned (off-resonant) cavity mode. In this situation, the dipole-dipole type of interaction can be realized as a consequence of the cavity photon exchange between the atoms. The dipole-dipole type of interaction, in turn, can be utilized as a controllable and deterministic entanglement mechanism, in which the cavity plays the role of a data bus that mediates this interaction. By considering this entanglement mechanism, in the second step we propose schemes for generation of the W entangled states for a chain of N three-level atoms, in which the atoms are equally distanced from each other and transported through the cavity by means of an optical lattice. The W state is a particular case of a Dicke state [64] and it is known as the *genuine* entangled state since it cannot be transformed into other entangled states under local operations and classical communication protocols [22]. Moreover, the properties of W states have been explored in details during recent years [23, 24] and it was found important for such practical applications like quantum teleportation, quantum dense coding, and quantum key distribution [25, 26, 27]. Various experiments have been reported in the literature for generation of three-qubit W states by using optical systems [28, 29], nuclear magnetic resonance [30], and ion trapping techniques [31].

In the first part of this manuscript, we describe our scheme to generate the state of W-class for N initially uncorrelated atomic qubits encoded in the chain of three-level atoms. This scheme works in a completely deterministic way and is based on the dipole-dipole interaction between distant atoms which are coupled simultaneously to an off-resonant optical cavity and a laser beam that acts perpendicularly to the cavity axis [63]. The two parameters that control this atom-cavity-laser interaction are (i) the velocity of atomic chain along the axis of lattice and (ii) the distance between the atoms. In the third part of this manuscript, furthermore, we determine the velocities and distances for which the initially uncorrelated atoms produce the W_N states for the chains consisting of $N = 2, 3, 4$ and 5 atoms. Apart from generation of the W states, we analyze how robust are the generated entangled states with respect to small oscillations in the atomic motion as caused by thermal effects. Finally, we discuss the implementation of our scheme by considering the setup similar to those utilized in the Institute of Applied Physics in Bonn [20], and we conclude that our schemes can be adapted to the near-future developments in cavity-QED.

Part I

Theory of the atom-light interaction

Chapter 1

Interaction of a two-level atom with a single-mode cavity field

On the first sight, the theoretical description of interaction between a single atom and the light field in the quantum regime is a formidable task since it involves many degrees of freedom associated with complicated atomic level structure. For the most of our purposes, however, it is justified to describe the atom as having only two electronic eigenstates, i.e., only two internal levels are involved. This *two-level* approximation is valid since we consider the interaction of an atom with a single-mode monochromatic light. In this case, the relevant atomic levels are those which satisfy the conditions: (i) the energy difference associated with atomic transition should match the energy of incident photon, and (ii) the selection rules do not inhibit the transition. This simplified model has been proposed and solved analytically by E. T. Jaynes and F. W. Cummings [13] in the early 1960s.

As we already mentioned, cavity-QED is a field of research which studies interaction of a single atom and electromagnetic field in confined spaces. The most interesting regime of atom-light interaction, moreover, is the (so-called) resonant interaction regime in which the above two conditions are satisfied. Experimentally, this regime is realized by placing a single atom inside the cavity with a very high quality of mirrors. Because of the mirrors quality and strong electric-dipole coupling between the atom and confined light field, the cavity photon can interact many times with the atom before it escapes. Theoretically, this regime provides an exceptional entanglement mechanism (see below) of atoms and cavity photons which are interpreted as bits of quantum information (qubits). From this perspective, therefore, cavity-QED offers an excellent framework for quantum information processing.

In this chapter, we first describe the mode structure of monochromatic light that propagates inside a planar cavity. We quantize the confined light and derive the Hamil-

CHAPTER 1: Interaction of a two-level atom with a single-mode cavity field

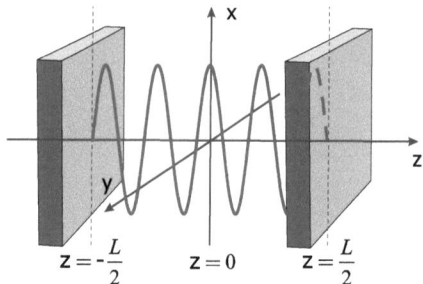

Figure 1.1: Planar (Fabry-Perot) cavity that consists of two reflecting mirrors and which are separated by distance L.

tonian that governs its evolution. Right afterwards, we introduce the coupling between a two-level atom and quantized light and derive the Jaynes-Cummings Hamiltonian that describes the model mentioned above. Moreover, we consider the situation in which both conditions (i) and (ii) are satisfied and derive the evolution of composite wavefunction associated with the coupled atom-cavity system. We find that this evolution describes a time-varying entanglement between a two-level atom and the cavity photon field. Finally, we analyze the effects of spontaneous atomic emission and cavity relaxation in order to understand how this time-varying entanglement evolves in realistic environments.

1.1 Quantized light field in a planar cavity

In this section we shall quantize the electromagnetic field inside a planar (Fabry-Perot) cavity that is displayed in Fig. 1.1. This cavity consists of two perfectly reflecting plane mirrors which are separated by an adjustable distance L. Specifically, we are interested in obtaining the energy of electromagnetic field and the electric field operator which, in the next sections, will be used to derive the interaction Hamiltonian for the coupled atom-cavity system.

Inside the cavity, the electromagnetic field fulfills the Maxwell's equations:

$$\nabla \cdot \mathbf{E}(\mathbf{r}, t) = 0, \tag{1.1a}$$

$$\nabla \cdot \mathbf{B}(\mathbf{r}, t) = 0, \tag{1.1b}$$

$$\nabla \times \mathbf{E}(\mathbf{r}, t) = -\frac{\partial \mathbf{B}(\mathbf{r}, t)}{\partial t}, \tag{1.1c}$$

$$\nabla \times \mathbf{B}(\mathbf{r}, t) = \frac{1}{c^2} \frac{\partial \mathbf{E}(\mathbf{r}, t)}{\partial t}, \tag{1.1d}$$

where $\mathbf{E}(\mathbf{r}, t)$, $\mathbf{B}(\mathbf{r}, t)$ are the vacuum electric and magnetic fields, respectively, and

1.1. Quantized light field in a planar cavity

the speed of light in vacuum $c = 1/\sqrt{\varepsilon_0 \mu_0}$ defines the electric permittivity ε_0 and the magnetic permeability μ_0. By taking the curl of Eq. (1.1c) and substituting it in Eq. (1.1d), we find the equation[1]

$$\nabla^2 \mathbf{E}(\mathbf{r},t) - \frac{1}{c^2}\frac{\partial^2 \mathbf{E}(\mathbf{r},t)}{\partial t^2} = 0, \quad (1.2)$$

which describes the propagation of electromagnetic field in the vacuum.

Perfectly reflecting mirrors of a planar cavity are separated from each other by the distance L such that the position of mirrors along the z-axis are given by $z_- = -L/2$ and $z_+ = +L/2$ as displayed in Fig. 1.1. These perfectly reflecting mirrors, moreover, imply the vanishing of tangential component of electric field and the normal component of magnetic field on the interior cavity sides

$$\mathbf{n}(\mathbf{r}_\pm) \times \mathbf{E}(\mathbf{r}_\pm,t) = 0, \quad \text{and} \quad \mathbf{n}(\mathbf{r}_\pm) \cdot \mathbf{B}(\mathbf{r}_\pm,t) = 0, \quad (1.3)$$

where $\mathbf{n}(\mathbf{r}_\pm)$ are the normal vectors at the positions $\mathbf{r}_\pm = (x,y,z_\pm)$ on the mirrors. Since the above boundary conditions are independent of time, it is possible to apply the separation ansatz $\mathbf{E}(\mathbf{r},t) = \mathcal{N}\, q(t)\, \mathbf{u}(\mathbf{r})$ [14] which substituted in Eq. (1.2), leads to

$$\frac{\nabla^2 \mathbf{u}(\mathbf{r})}{\mathbf{u}(\mathbf{r})} = \frac{1}{c^2}\frac{\ddot{q}(t)}{q(t)}, \quad (1.4)$$

where dot denotes the time derivative and $\mathcal{N}$ denotes the normalization factor. Instead of Eq. (1.4), equivalently, we can consider the system of two equations[2]

$$\nabla^2 \mathbf{u}(\mathbf{r}) + k^2\, \mathbf{u}(\mathbf{r}) = 0, \quad (1.5a)$$
$$\ddot{q}(t) + \omega^2\, q(t) = 0, \quad (1.5b)$$

where $k = |\mathbf{k}|$ is the modulus of wave vector and $\omega = c\,k$ is the wave frequency determined by the wave vector.

The first equation from (1.5) is the Helmholtz equation which describes the cavity field structure, while the second is equation for the classical harmonic oscillator. In order to simplify further derivations, below, we consider the ansatz for the field structure

$$\mathbf{u}(\mathbf{r}) = \mathbf{x}\cos(k\,z), \quad (1.6)$$

which describes a standing wave that propagates between the mirrors and is linearly polarized in the x direction [see Fig. 1.1]. It can be readily checked that the function (1.6) fulfills the Helmholtz equation (1.5a) and the boundary conditions (1.3) only for

---

[1] In order to obtain this equation, the identity $\nabla \times \nabla \mathbf{E}(\mathbf{r},t) = \nabla(\nabla \cdot \mathbf{E}(\mathbf{r},t)) - \nabla^2 \mathbf{E}(\mathbf{r},t)$ along with Eq. (1.1a) have been used.

[2] Note that we could choose here the negative sign instead of positive. However, this choice would lead to the exponentially growing solutions for $q(t)$, which are nor physically acceptable.

the discrete values $k_\ell = (2\ell + 1)\pi/L$, where ℓ is an integer. These discrete values imply, in turn, the discrete set of functions $\mathbf{u}_\ell(\mathbf{r})$ and the discrete wave frequencies $\omega_\ell = c(2\ell + 1)\pi/L$ which are supported by the cavity. Any two functions $\mathbf{u}_\ell(\mathbf{r})$ and $\mathbf{u}_{\ell'}(\mathbf{r})$, moreover, satisfy the orthogonality condition

$$\frac{2}{L} \int_{z_-}^{z_+} \mathbf{u}_\ell(\mathbf{r}) \cdot \mathbf{u}_{\ell'}(\mathbf{r})\, dz = \delta_{\ell,\ell'}. \tag{1.7}$$

Recall that due to the separation ansatz that we have applied above, the electric field

$$\mathbf{E}_c(\mathbf{r}, t) = \mathcal{N}\, q_c(t)\, \mathbf{x}\, \cos(k_c z), \tag{1.8}$$

is given by both the field structure $\mathbf{u}_\ell(\mathbf{r})$ and the oscillator $q(t)$ that satisfies Eq. (1.5b). Moreover, we have taken only one single mode (from the discrete set of modes) that corresponds to the wave vector modulus k_c and frequency ω_c. Substituting expression (1.8) into Eq. (1.1c) and using the oscillator equation (1.5b), we find the expression for the magnetic field

$$\mathbf{B}_c(\mathbf{r}, t) = -\int \frac{\mathcal{N}}{c^2\, k_c}\, \mathbf{y}\, \ddot{q}_c(t)\, \sin(k_c z)\, dt = -\frac{\mathcal{N}}{c^2\, k_c}\, \mathbf{y}\, \dot{q}_c(t)\, \sin(k_c z). \tag{1.9}$$

Having both the electrical and magnetic components, we can derive the field Hamiltonian

$$H_c = \frac{1}{2} \int_{z_-}^{z_+} \left[\varepsilon_0\, \mathbf{E}_c^2(\mathbf{r}, t) + \mathbf{B}_c^2(\mathbf{r}, t)/\mu_0\right] dx = \frac{1}{2}\left[\omega_c^2\, q_c^2(t) + p_c^2(t)\right], \tag{1.10}$$

where $p_c(t) \equiv \dot{q}_c(t)$ and the normalization factor $\mathcal{N} = \sqrt{\frac{2\omega_c^2}{\varepsilon_0 L}}$ has been introduced. Hamiltonian (1.10) is formally identical with the Hamiltonian for a classical harmonic oscillator of unit mass that oscillates with the frequency ω_c, and where $q(t)$ and $p(t)$ are the position and momentum of oscillator. Below, we shall omit for brevity all subscripts and introduce the dimensionless complex amplitude $a(t) = \frac{p(t) - i\,\omega\, q(t)}{\sqrt{2\hbar\omega}}$, where $\hbar$ is the Planck's constant. The oscillator equation (1.5b) implies the equation for the complex amplitude $\dot{a}(t) = -i\,\omega\, a(t)$ with the general solution $a(t) = a\, e^{-i\omega t}$, where $a \equiv a(0)$. By substituting $p(t)$ and $q(t)$ together with solutions for $a(t)$ into the Hamiltonian (1.10), we find the energy of one single (oscillation) mode that corresponds to the frequency ω

$$H = \frac{1}{2}\hbar\omega\, (a^*\, a + a\, a^*). \tag{1.11}$$

The quantization of harmonic oscillator (1.10) is done by postulating the equal-time commutation relations: $[\hat{q}(t), \hat{p}(t)] = i\hbar$ and $[\hat{q}(t), \hat{q}(t)] = [\hat{p}(t), \hat{p}(t)] = 0$, where $\hat{q}(t)$ and $\hat{p}(t)$ become the position and momentum operators of oscillator. This procedure implies, therefore, that the complex amplitudes a and a^* become the operators $\hat{a}$ and $\hat{a}^\dagger$, respectively, which satisfy the commutation relations

$$[\hat{a}, \hat{a}^\dagger] = 1, \quad \text{and} \quad [\hat{a}, \hat{a}] = 0. \tag{1.12}$$

1.1. Quantized light field in a planar cavity

By substituting the operators $\hat{a}$ and $\hat{a}^\dagger$ in the Hamiltonian (1.11) and using the commutation relations (1.12), we find the quantized energy of a single-mode field that corresponds to the frequency ω

$$\hat{H} = \frac{1}{2}\hbar\omega\left(\hat{a}^\dagger\hat{a} + \hat{a}\hat{a}^\dagger\right) = \hbar\omega\left(\hat{a}^\dagger\hat{a} + \frac{1}{2}\hat{1}\right). \tag{1.13}$$

The same procedure applied to the electric field (1.8) yields the operator

$$\hat{\mathbf{E}}(\mathbf{r},t) = i\sqrt{\frac{\hbar\omega}{\varepsilon_0 L}}\left(\hat{a}\,e^{-i\omega t} - \hat{a}^\dagger\,e^{i\omega t}\right)\mathbf{x}\cos(k\,z), \tag{1.14}$$

which describes the single-mode quantized intracavity electric field.

Note that the quantization of electromagnetic field arises solely due to the time-dependent part $q(t)$ of electric field component, while the spatial part $\mathbf{u}(\mathbf{r})$ is purely classical which satisfies the Helmholtz equation together with the boundary and orthogonality conditions. The results of Ref. [15] ensure, moreover, that the Helmholtz equation for any finite cavity shape has a complete and orthogonal set of eigenfunctions $\mathbf{u}_\ell(\mathbf{r})$ which together with the eigenvalues k_ℓ are completely determined by the cavity shape and replace the combination $\{\mathbf{u}_\ell(\mathbf{r}) = \mathbf{x}\cos(k_\ell z),\, k_\ell = (2\ell+1)\pi/L\}$ that we found for the planar (Fabry-Perot) cavity. The quantization procedure we performed, therefore, remains valid for any finite cavity with arbitrary shape and in order to summarize the obtained results in a form that is independent on the cavity shape, we express the electric field (1.14) in the form

$$\hat{\mathbf{E}}(\mathbf{r},t) = i\sqrt{\frac{\hbar\omega}{2\varepsilon_0 \mathrm{V}}}\left(\hat{a}\,e^{-i\omega t} - \hat{a}^\dagger\,e^{i\omega t}\right)\boldsymbol{\epsilon}\,f(\mathbf{r}), \tag{1.15}$$

where $\boldsymbol{\epsilon}$ is a real unit vector that denotes the field polarization, $f(\mathbf{r})$ is a real scalar function that describes the field structure, and $\mathrm{V} = \int |f(\mathbf{r})|^2\,d^3\mathbf{r}$ denotes the cavity mode volume.

The structure of Hilbert space associated with operators $\hat{a}$ and $\hat{a}^\dagger$ is determined by the positivity of $\hat{H}$ and, therefore, by the set of its positive eigenvalues E_n. It can be shown that

$$\hat{H}\,\hat{a}^\dagger\,|\phi_n\rangle = (E_n + \hbar\omega)\,\hat{a}^\dagger\,|\phi_n\rangle \quad\text{and}\quad \hat{H}\,\hat{a}\,|\phi_n\rangle = (E_n - \hbar\omega)\,\hat{a}\,|\phi_n\rangle, \tag{1.16}$$

which implies that $\hat{a}\,|\psi_n\rangle$ is an eigenstate of $\hat{H}$ with the decreased energy $(E_n - \hbar\omega)$, while $\hat{a}^\dagger\,|\phi_n\rangle$ is an eigenstate of $\hat{H}$ with the increased energy $(E_n + \hbar\omega)$. The operator $\hat{a}$ is called the annihilation operator since it decreases E_n exactly by the energy of one single photon $\hbar\omega$. In contrast, the operator $\hat{a}^\dagger$ is called the creation operator since it increases E_n by the energy of one single photon. The positivity of $\hat{H}$ implies, moreover, that the field energy cannot be decreased infinitely and, therefore, the Hilbert space must

CHAPTER 1: Interaction of a two-level atom with a single-mode cavity field

include the ground state $|0\rangle$ which satisfies the condition $\hat{a}|0\rangle = 0$. This ground state, therefore, is called the vacuum state since it contains no photons and it corresponds to the lowest energy $\hbar\omega/2$ of light field.

Starting from the ground state, moreover, any quanta of energy (photons) can be produced by multiple application of the (normalized) creation operator

$$|n\rangle = \frac{1}{\sqrt{n!}} \left(\hat{a}^\dagger\right)^n |0\rangle, \tag{1.17}$$

where $|n\rangle$ denotes the number (Fock) state which is defined as the eigenstate of $\hat{H}$ that corresponds to the eigenvalue $\hbar\omega(n + 1/2)$. The relation (1.17) implies the action of operators $\hat{a}$ and $\hat{a}^\dagger$ on the Fock states

$$\hat{a}|n\rangle = \sqrt{n}|n-1\rangle, \quad \hat{a}^\dagger|n\rangle = \sqrt{n+1}|n+1\rangle \tag{1.18}$$

and also ensures that the Fock states form an orthogonal basis of the Hilbert space in question

$$\langle n|n'\rangle = \delta_{n,n'}, \quad \sum_{n=0}^{\infty} |n\rangle\langle n| = \hat{I}. \tag{1.19}$$

We conclude, therefore, that the Hilbert space associated with operators $\hat{a}$ and $\hat{a}^\dagger$ is spanned by the Fock states and, further, we shall solely consider the Fock states in order to characterize a quantum state of cavity field.

1.1.1 Transverse cavity field components

So far, we assumed that the cavity field structure $\mathbf{u}(\mathbf{r}) = \boldsymbol{\epsilon} f(\mathbf{r})$ is determined by the polarization $\boldsymbol{\epsilon}$ and the real function $f(\mathbf{r}) = \cos(k\,z)$ which describes a standing wave that propagates between two planar mirrors. This scalar field fulfills the Helmholtz equation and the boundary conditions (1.3) for discrete values $k_\ell = (2\,\ell + 1)\,\pi/L$ which imply, in turn, discrete wave frequencies $\omega_\ell = c\,(2\,\ell + 1)\,\pi/L$ supported by the cavity. Below, we shall demonstrate that the light field of a cavity that consists of two spherical mirrors can support two transverse components if the radii of mirrors coincide with the radii of curvature of light wavefront.

In order to show this, we shall consider the following ansatz

$$f(\mathbf{r}) = \mathrm{Re}\left[F(\mathbf{r})\, e^{-i\,k z}\right], \tag{1.20}$$

where

$$F(\mathbf{r}) = e^{-i\,Q(z)(x^2+y^2)}\, e^{-i\,P(z)}. \tag{1.21}$$

By substituting the ansatz (1.20) into the Helmholtz equation, we obtain

$$\frac{\partial^2 F(\mathbf{r})}{\partial x^2} + \frac{\partial^2 F(\mathbf{r})}{\partial y^2} + \frac{\partial^2 F(\mathbf{r})}{\partial z^2} - 2\,i\,k\,\frac{\partial F(\mathbf{r})}{\partial z} = 0. \tag{1.22}$$

1.1. Quantized light field in a planar cavity

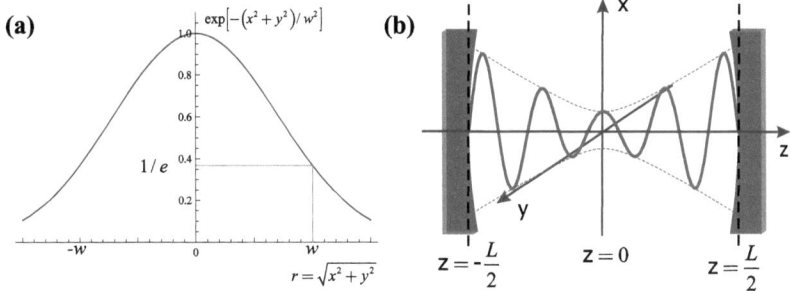

Figure 1.2: (a) The cavity waist w is defined as the half-width of the Gaussian function $\exp[-r^2/w^2]$ at the point where the amplitude is $1/e$ of its maximum value. (b) Cavity that consists of two spherical mirrors which are separated by the distance L.

Notice that the exponent e^{-ikz}, which describes the standing wave propagating along the z-axis, is factored out and the remaining z dependence of $F(\mathbf{r})$ is caused basically by the diffraction effects. Therefore, this z dependence of $F(\mathbf{r})$ must be slow if compared to the optical wavelength k and also if compared to the transverse variations of $F(\mathbf{r})$. This assumption, which is know as the *paraxial* approximation, allows us to express (1.22) in the form

$$\frac{\partial^2 F(\mathbf{r})}{\partial x^2} + \frac{\partial^2 F(\mathbf{r})}{\partial y^2} - 2ik\frac{\partial F(\mathbf{r})}{\partial z} = 0. \tag{1.23}$$

According to the ansatz (1.21), moreover, we explicitly assumed that the transverse field is characterized by the exponent which depends on $(x^2 + y^2)$. The light wave that admits such dependence on the transversal coordinates and is the solution of Eq. (1.23) is called the Gaussian beam and it exhibits one minimal diffractive spreading of the light [66].

By substituting the ansatz (1.21) into the paraxial equation (1.23), we obtain the equations

$$k\frac{dP(z)}{dz} + 2iQ(z) = 0, \quad 2Q(z)^2 + k\frac{dQ(z)}{dz} = 0, \tag{1.24}$$

which being solved and substituted back into (1.21) gives [67]

$$F(\mathbf{r}) = \frac{w}{W(z)} \exp\left[-\frac{x^2+y^2}{W(z)^2}\right] \exp\left[-i\frac{k(r^2+y^2)}{2R(z)}\right] \exp\left[i \arctan\left(\frac{\lambda z}{\pi w^2}\right)\right], \tag{1.25}$$

where

$$W(z) = w\sqrt{1 + \left(\frac{\lambda z}{\pi w^2}\right)^2} \quad \text{and} \quad R(z) = z\left[1 + \left(\frac{\pi w^2}{\lambda z}\right)^2\right] \tag{1.26}$$

denote the width of transverse Gaussian profile and the radius of curvature of light wavefront, respectively. In the above notation, moreover, we introduced the (so-called)

CHAPTER 1: Interaction of a two-level atom with a single-mode cavity field

beam waist w which is equal to the half-width of the Gaussian function at the point where the amplitude is $1/e$ of its maximum value [see Fig. 1.2(a)]. The first exponent in (1.25) describes the amplitude distribution across the light wave, the second term describes the structure of nearly-plane wave (paraxial-spherical wave), and the third term is the Guoy phase that describes the longitudinal phase delay of the beam.

Owning to the solution (1.25) of paraxial equation (1.23), we substitute it into the ansatz (1.20) and obtain the expression for the cavity field structure

$$f(\mathbf{r}) = \frac{w}{W(z)} \exp\left[-\frac{(x^2+y^2)}{W(z)^2}\right] \cos\left[k\,z - \frac{k\,(x^2+y^2)}{2\,R(z)} + G(z)\right], \quad (1.27)$$

where $G(z) = \arctan[\lambda\,z/(\pi\,w^2)]$ denotes the Guoy phase. It is clear, however, that the Gaussian beam (1.27) will not form a standing wave inside a planar (Fabry-Perot) cavity. The reason for this is that the wavefront of a Gaussian beam is not a plane wave but rather a nearly-plane wave (paraxial-spherical wave) with the radius of curvature being characterized by $R(z)$. Therefore, instead of the planar cavity, we shall consider a cavity with two spherical mirrors characterized by the same (constant) radii of curvature R_M and which are separated by an adjustable distance L (in the origin of $x-y$ plane) as displayed in Fig. 1.2(b). Therefore, if the radii of curvature of mirrors R_M are equal to the radii of curvature of wavefront $R(z)$ taken at $z = z_\pm$, then the beam is reflected back into itself and a standing wave can be produced.

Finally, we need to calculate the set of frequencies ω_ℓ supported by the cavity and which should replace the set $c\,(2\,\ell+1)\,\pi/L$ found by us for a planar cavity. In order to proceed, notice that the boundary conditions (1.3) imply that along the z-axis (in the origin of $x-y$ plane) the cavity can accommodate $n + 1/2$ wavelengths and, therefore, the cosine argument of (1.27) must fulfill the condition

$$\left(k_\ell\,z_+ - \frac{k\,(x^2+y^2)}{2\,R(z_+)} + G(z_+)\right) - \left(k_\ell\,z_- - \frac{k\,(x^2+y^2)}{2\,R(z_-)} + G(z_-)\right) = \pi\,(2\,\ell+1). \quad (1.28)$$

Since $R_M = R(z_+) = R(z_-)$, the above condition becomes $k_\ell\,L + G(z_+) - G(z_-) = \pi\,(2\,\ell+1)$ and yields the set of wave numbers $k_\ell = [\pi\,(2\,\ell+1) + G(z_-) - G(z_+)]/L$ which are compatible with the cavity boundary conditions. Owning to this set of wave numbers, we readily determine the set of frequencies

$$\omega_\ell = c\,\frac{\pi\,(2\,\ell+1) + G(z_-) - G(z_+)}{L} \quad (1.29)$$

which are supported by the cavity and which, in contrast to the planar cavity, involve the difference between the Guoy phases taken at the cavity walls.

1.2 Atom coupled to a cavity

In the previous section we obtained the expressions (1.13) and (1.15) for energy and electromagnetic field which evolve freely inside a planar or spherical cavity. In this

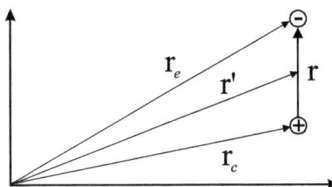

Figure 1.3: The model of hydrogen-like atom consisting of a hard-core at position $\mathbf{r}_c$ and one valence electron at position $\mathbf{r}_e$. Because of non-vanishing mass ratio between electron and atomic core, the center-of-mass position $\mathbf{r}'$ is shifted. See the text for details.

section we derive the Hamiltonian that governs the interaction of cavity field coupled to a two-level atom.

We consider an atom at rest being placed inside a cavity which is characterized by the cavity field structure $f(\mathbf{r})$ and frequency $\omega = kc$. We assume that the atom consists of electrons surrounding its nucleus which form together a hard-core with one valence electron. If we denote by $\mathbf{r}_c$ and $\mathbf{r}_e$ the positions of hard-core and valence electron, respectively, then the interaction between atom and the intracavity field is governed by the dipole-field Hamiltonian

$$\hat{H}_{\text{int}} = -q\,\hat{\mathbf{r}} \cdot \hat{\mathbf{E}}(\mathbf{r}', t), \tag{1.30}$$

where q is the electric charge, $\hat{\mathbf{E}}(\mathbf{r}, t)$ is the electric field operator (1.15), $\mathbf{r} = \mathbf{r}_e - \mathbf{r}_c$ denotes the relative position of valence electron, and $\mathbf{r}' = (m_e\,\mathbf{r}_e + m_c\,\mathbf{r}_c)/(m_e + m_c)$ is the atomic center-of-mass position [see Fig. 1.3]. Note that in the above Hamiltonian, we assumed that the electric field depends on the position of atomic center-of-mass $\mathbf{r}'$. This physically justified assumption, which is known as the *dipole* approximation, is valid whenever the wavelength of light field is large in comparison with the size of atom and implies that the variation of electric field over the atom can be neglected.

The internal structure of a two-level atom is completely characterized by the states $|g\rangle$ (ground) and $|e\rangle$ (excited) which are accessible to the valence electron and which correspond to the energies $E_g = \hbar\omega_g$ and $E_e = \hbar\omega_e$, respectively ($E_g < E_e$). The states $|g\rangle$ and $|e\rangle$, moreover, fulfill the orthogonality and completeness relations

$$\langle e|g\rangle = \langle g|e\rangle = 0, \quad \langle g|g\rangle = \langle e|e\rangle = 1, \quad |e\rangle\langle e| + |g\rangle\langle g| = \hat{\mathbf{I}} \tag{1.31}$$

and they are the eigenstates of atomic Hamiltonian

$$\hat{H}_{\text{a}} = E_g\,|g\rangle\langle g| + E_e\,|e\rangle\langle e| = \frac{1}{2}\hbar\,(\omega_e + \omega_g)\,\hat{\mathbf{I}} + \hbar\,\omega_a\,\hat{\sigma}^z, \tag{1.32}$$

where $\omega_a = \omega_e - \omega_g$ is the atomic transition frequency and $\hat{\sigma}^z = \frac{1}{2}(|e\rangle\langle e| - |g\rangle\langle g|)$. We use relations (1.31) to express the Hamiltonian $\hat{H}_{\text{a-f}} = \hat{H}_{\text{a}} + \hat{H}_{\text{int}}$, which describes a

CHAPTER 1: Interaction of a two-level atom with a single-mode cavity field

two-level atom coupled to the intracavity field, in the form

$$\hat{H}_{\text{a-f}} = \hat{I}\left[\hat{H}_a - q\,\hat{\mathbf{r}} \cdot \hat{\mathbf{E}}(\mathbf{r}',t)\right]\hat{I} = \hat{H}_a - d\left(\hat{\sigma}^\dagger \boldsymbol{\epsilon}_a^* + \hat{\sigma}\,\boldsymbol{\epsilon}_a\right) \cdot \hat{\mathbf{E}}(\mathbf{r}',t), \qquad (1.33)$$

where $\hat{\sigma} = |g\rangle\langle e|$ and $\hat{\sigma}^\dagger = |e\rangle\langle g|$ denote the excitation lowering and rising operators, respectively, and where we introduced the notation $\langle g|q\,\hat{\mathbf{r}}|e\rangle = d\,\boldsymbol{\epsilon}_a$ with d being the real dipole matrix element of the atomic transition[3] and $\boldsymbol{\epsilon}_a$ being the unit complex vector that determines the atomic transition polarization.

In order to simplify our derivations, we need to specify the atomic transition polarization $\boldsymbol{\epsilon}_a$. The simplified picture of two-level atom that we have considered above, however, does not tell anything about the polarization of light that is emitted or absorbed during the atomic transition. In any realistic hydrogen-like system, moreover, each atomic level $|g\rangle$ or $|e\rangle$ consists of a manifold of almost degenerate sub-levels, which are labeled by the quantum numbers m_e and m_g, respectively. These quantum numbers give rise to the revised dipole matrix element $\langle g, m_g|q\,\hat{\mathbf{r}}|e, m_e\rangle$ in which the transition polarization is determined by the values of $\Delta m = m_g - m_e = \{-1, 0, +1\}$. It can be shown [94], that the transitions with $\Delta m = +1$ and $\Delta m = -1$ produce the circular polarizations $\boldsymbol{\epsilon}_a^+ = \frac{1}{\sqrt{2}}(\mathbf{x} + i\,\mathbf{y})$ and $\boldsymbol{\epsilon}_a^- = \frac{1}{\sqrt{2}}(\mathbf{x} - i\,\mathbf{y})$, respectively, while the transition with $\Delta m = 0$ produces the polarization $\boldsymbol{\epsilon}_a^\pi = \mathbf{z}$.

In the second part of this manuscript, we show that the atoms which are utilized in the typical cavity-QED experiments support the circular polarization $\boldsymbol{\epsilon}_a^+$ for the dipole matrix element associated with the $|g\rangle \to |e\rangle$ transition. Placed inside the cavity, therefore, such an atom couples to the cavity electric field

$$\hat{\mathbf{E}}(\mathbf{r},t) = i\,f(\mathbf{r})\left[\sqrt{\frac{\hbar\omega}{2\varepsilon_0 V}}\left(\hat{a}\,e^{-i\omega t} - \hat{a}^\dagger e^{i\omega t}\right)\mathbf{x} + \sqrt{\frac{\hbar\widetilde{\omega}}{2\varepsilon_0 V}}\left(\hat{b}\,e^{-i\widetilde{\omega} t} - \hat{b}^\dagger e^{i\widetilde{\omega} t}\right)\mathbf{y}\right], \qquad (1.34)$$

where

$$[\hat{b}, \hat{b}^\dagger] = 1, \quad \text{and} \quad [\hat{b},\hat{b}] = [\hat{a},\hat{b}] = [\hat{a},\hat{b}^\dagger] = 0 \qquad (1.35)$$

and which involves both (orthogonally polarized) cavity modes in contrast to (1.15). The cavity mode that is polarized along the x-axis (C_x mode) is characterized by the operators $\{\hat{a}, \hat{a}^\dagger\}$ and frequency ω, while the orthogonal mode that is polarized along the y-axis (C_y mode) is characterized by the operators $\{\hat{b}, \hat{b}^\dagger\}$ along with frequency $\widetilde{\omega}$. Notice that a perfectly spherical shape of cavity mirrors implies that both cavity mode are strictly degenerate, i.e., $\omega = \widetilde{\omega}$. However, due to various imperfections caused by the fabrication process, a birefringent splitting between the orthogonally polarized cavity modes is produced and it makes one of the modes to be detuned with respect to

---

[3]In the expression (1.33), we used the fact that the diagonal matrix elements $\langle e|\hat{\mathbf{r}}|e\rangle$ and $\langle g|\hat{\mathbf{r}}|g\rangle$ of dipole operator vanish. This result can be readily demonstrated by considering the space representation of the diagonal matrix elements and noticing that the complete integrand is anti-symmetric and, therefore, vanishes.

another, i.e., $\delta = \omega - \widetilde{\omega} > 0$. Notice that since the cavity frequency ω is usually very big if compared to the birefringent splitting δ, we can safely replace $\widetilde{\omega}$ by ω in the terms where this frequency appears linearly

$$\sqrt{\frac{\hbar\widetilde{\omega}}{2\varepsilon_0 V}} \cong \sqrt{\frac{\hbar\omega}{2\varepsilon_0 V}}. \tag{1.36}$$

By inserting the electric field operator (1.34) together with polarization $\boldsymbol{\epsilon}_a = \boldsymbol{\epsilon}_a^+$ and (1.36) in the Hamiltonian (1.33), we find

$$\begin{aligned}\hat{H}_{\text{a-f}} = \hat{H}_{\text{a}} &- i\frac{\hbar}{2}\hat{\sigma}^\dagger \left[\left(\hat{a}e^{-i\omega t} - \hat{a}^\dagger e^{i\omega t}\right) - i\left(\hat{b}e^{-i\widetilde{\omega} t} - \hat{b}^\dagger e^{i\widetilde{\omega} t}\right)\right] g(\mathbf{r}') \\ &- i\frac{\hbar}{2}\hat{\sigma}\left[\left(\hat{a}e^{-i\omega t} - \hat{a}^\dagger e^{i\omega t}\right) + i\left(\hat{b}e^{-i\widetilde{\omega} t} - \hat{b}^\dagger e^{i\widetilde{\omega} t}\right)\right] g(\mathbf{r}'),\end{aligned} \tag{1.37}$$

where we introduced the interaction frequency (atom-cavity coupling)

$$g(\mathbf{r}) = g_\circ\, f(\mathbf{r}) = d\sqrt{\frac{\omega}{\varepsilon_0 \hbar V}}\, f(\mathbf{r}) \tag{1.38}$$

between the atom and cavity modes. Notice also that the atom-cavity coupling is determined entirely by the atomic dipole momentum, mode frequency, cavity field amplitude, and the effective mode volume.

Hamiltonian (1.37) is further simplified by switching to the interaction picture with respect to the time-independent part $\hat{H}_a$, or equivalently, the time-dependent parts of (1.37) is transformed by means of the unitary operator $\hat{U}_a = \exp\left(-\frac{i}{\hbar}\hat{H}_a t\right)$

$$\begin{aligned}\hat{H}_I &= \hat{U}_a^\dagger \left(\hat{H}_{\text{a-f}} - \hat{H}_a\right)\hat{U}_a \\ &= -i\frac{\hbar}{2}g(\mathbf{r}')\left(\hat{\sigma}^\dagger \hat{a}\, e^{i\Delta t} - \hat{\sigma}\, \hat{a}^\dagger e^{-i\Delta t}\right) - \frac{\hbar}{2}g(\mathbf{r}')\left(\hat{\sigma}^\dagger \hat{b}\, e^{i(\Delta+\delta)t} + \hat{\sigma}\hat{b}^\dagger e^{-i(\Delta+\delta)t}\right),\end{aligned} \tag{1.39}$$

where $\Delta = \omega_a - \omega$ is the difference between the atomic transition frequency and the field frequency of mode C_x, to which we shall refer as the atom-cavity detuning. Note that in the Hamiltonian (1.39), we neglected the terms $\hat{\sigma}\,\hat{a}\,e^{-i(\omega_a+\omega)t}$, $\hat{\sigma}^\dagger\hat{a}^\dagger e^{i(\omega_a+\omega)t}$, $\hat{\sigma}\,\hat{b}\,e^{-i(\omega_a+\widetilde{\omega})t}$, and $\hat{\sigma}^\dagger\hat{b}^\dagger e^{i(\omega_a+\widetilde{\omega})t}$ since the contribution of these fast oscillating terms is averaged much rapidly if compared to the slow oscillating terms and, therefore, these fast oscillating terms play a minor role in the evolution of coupled atom-cavity system[4].

1.2.1 Resonant atom-cavity interaction

Hamiltonian (1.30) governs the interaction of a two-level atom with two (non-degenerate) orthogonal modes of quantized light field. The physically interesting regime of atom-cavity evolutions is the so-called resonant interaction regime, in which the atomic transition frequency ω_a matches the frequency of one of cavity modes, i.e., $\omega_a = \omega$ or $\omega_a = \widetilde{\omega}$.

---
[4]This assumption, which is known as the *rotating wave* approximation, is valid whenever the applied electromagnetic field is nearly resonant with the atomic transition.

CHAPTER 1: Interaction of a two-level atom with a single-mode cavity field

Let us assume that the atom is tuned in resonance with mode C_x such that the detuning Δ vanishes and the Hamiltonian (1.39) becomes

$$\hat{H}_I = -i\,\hbar\,\frac{g_\circ}{2}\left(\hat{\sigma}^\dagger \hat{a} - \hat{\sigma}\,\hat{a}^\dagger\right) - \hbar\,\frac{g_\circ}{2}\left(\hat{\sigma}^\dagger \hat{b}\,e^{i\delta t} + \hat{\sigma}\,\hat{b}^\dagger\,e^{-i\delta t}\right), \tag{1.40}$$

where we also assumed that the atomic center-of-mass is located at the point $\mathbf{r}' = \mathbf{r}'_\circ$, in which the cavity field amplitude is maximum, i.e., $f(\mathbf{r}'_\circ) = 1$.

It is relevant to point out that for a large enough birefringent splitting δ, we can safely neglect the second term in the Hamiltonian (1.40) and consider the atom interacting only with the mode C_x. In order to show this, we expand in series the evolution operator associated with Hamiltonian (1.40) and keep the terms up to the second order. Then, by performing the integration and retaining only linear in time contributions, we express this evolution operator in the form

$$\hat{U}_I \cong \hat{\mathbf{1}} - \frac{i}{\hbar}\int_0^t \hat{H}_I\,dt' - \frac{1}{\hbar^2}\int_0^t\left(\hat{H}_I \int_0^{t'}\hat{H}_I\,dt''\right)dt' \cong \hat{\mathbf{1}} - \frac{i}{\hbar}\hat{H}_I^{\text{eff}}\,t \cong \exp\left[-\frac{i}{\hbar}\hat{H}_I^{\text{eff}}\,t\right], \tag{1.41}$$

where the effective Hamiltonian is given by

$$\hat{H}_I^{\text{eff}} = -i\,\hbar\,\frac{g_\circ}{2}\left(\hat{\sigma}^\dagger \hat{a} - \hat{\sigma}\,\hat{a}^\dagger\right) - \hbar\,\frac{g_\circ^2}{4\delta}\left(\hat{\sigma}\hat{\sigma}^\dagger\,\hat{b}^\dagger \hat{b} - \hat{\sigma}^\dagger\hat{\sigma}\,\hat{b}\,\hat{b}^\dagger\right). \tag{1.42}$$

The second term of this (effective) Hamiltonian produces only the phase shifts

$$|e;\bar{n}\rangle \to \exp\left[-i\,\frac{g_\circ^2\,(\bar{n}+1)}{4\delta}\right]|e;\bar{n}\rangle \quad \text{and} \quad |g;\bar{n}\rangle \to \exp\left[i\,\frac{g_\circ^2\,\bar{n}}{4\delta}\right]|g;\bar{n}\rangle, \tag{1.43}$$

which can be neglected if the condition $\delta \gg g_\circ$ is satisfied. In other words, the resonant interaction between a circularly polarized atom and both (orthogonally polarized) cavity modes reduces to the resonant interaction between the atom and mode C_x, whenever $\Delta = 0$ and the birefringent splitting δ is sufficiently detuned with respect to the atom-cavity coupling $g_\circ$. Therefore, we can safely describe the atom-cavity evolution by using only the first term of the effective Hamiltonian (1.42)

$$\hat{H}_x = -i\,\hbar\,\frac{g_\circ}{2}\left(\hat{\sigma}^\dagger \hat{a} - \hat{\sigma}\,\hat{a}^\dagger\right). \tag{1.44}$$

By a similar line of reasoning, furthermore, it can be shown that by adjusting (tuning) the atomic transition in resonance with mode C_y, i.e., $\Delta = -\delta$, the atom-cavity evolution is governed by the Hamiltonian

$$\hat{H}_y = -\hbar\,\frac{g_\circ}{2}\left(\hat{\sigma}^\dagger\,\hat{b} + \hat{\sigma}\,\hat{b}^\dagger\right), \tag{1.45}$$

and where the same condition, i.e., $\delta \gg g_\circ$, must hold true. Hamiltonian of the type (1.44) or (1.45) was originally introduced and analyzed by E. T. Jaynes and F. W. Cummings in Ref. [13] and is known in the literature as the Jaynes-Cummings Hamiltonian.

In order to solve the Schrödinger equations associated with Hamiltonians (1.44) and (1.45)

$$i\hbar\frac{d|\Psi_x(t)\rangle}{dt} = \hat{H}_x|\Psi_x(t)\rangle, \quad i\hbar\frac{d|\Psi_y(t)\rangle}{dt} = \hat{H}_y|\Psi_y(t)\rangle, \tag{1.46}$$

we assume the ansatz for the wave-functions

$$|\Psi_x(t)\rangle = \sum_{n=0}^{\infty}\left[\Psi_x^{e,n}(t)|e;n\rangle + \Psi_x^{g,n+1}(t)|g;n+1\rangle\right] + \Psi_x^{g,0}(t)|g;0\rangle, \tag{1.47a}$$

$$|\Psi_y(t)\rangle = \sum_{n=0}^{\infty}\left[\Psi_y^{e,\bar{n}}(t)|e;\bar{n}\rangle + \Psi_y^{g,\bar{n}+1}(t)|g;\bar{n}+1\rangle\right] + \Psi_y^{g,\bar{0}}(t)|g;\bar{0}\rangle, \tag{1.47b}$$

where the states $|n\rangle$ and $|\bar{n}\rangle$ refer to the Fock spaces associated with the modes C_x and C_y, respectively. By substituting this ansatz into (1.46), we obtain two closed systems of equations

$$\dot{\Psi}_x^{e,n}(t) = -\frac{g_\circ\sqrt{n+1}}{2}\Psi_x^{g,n+1}(t); \quad \dot{\Psi}_x^{g,n+1}(t) = \frac{g_\circ\sqrt{n+1}}{2}\Psi_x^{e,n}(t); \quad \dot{\Psi}_x^{g,0}(t) = 0, \tag{1.48a}$$

$$\dot{\Psi}_y^{e,\bar{n}}(t) = i\frac{g_\circ\sqrt{\bar{n}+1}}{2}\Psi_y^{g,\bar{n}+1}(t); \quad \dot{\Psi}_y^{g,\bar{n}+1}(t) = i\frac{g_\circ\sqrt{\bar{n}+1}}{2}\Psi_y^{e,\bar{n}}(t); \quad \dot{\Psi}_y^{g,\bar{0}}(t) = 0, \tag{1.48b}$$

where dot denotes the time derivative.

Equations (1.48) admit analytical solutions which together with the initial states $|\Psi_x(0)\rangle$ and $|\Psi_y(0)\rangle$ determine the time-evolution of wave-functions (1.47). We readily recognize, moreover, that the evolution of states $|g;0\rangle$ and $|g;\bar{0}\rangle$ is decoupled from the atom-cavity dynamics and the wave-functions $|\Psi_x(t)\rangle$ or $|\Psi_y(t)\rangle$ remain trapped for the overall interaction time if $|\Psi_x(0)\rangle = |g;0\rangle$ or $|\Psi_y(0)\rangle = |g;\bar{0}\rangle$, respectively. We are interested, however, in the atom-cavity evolution obtained for two initial states: (i) cavity is empty and the atom is excited, i.e., $|\Psi_x^e(0)\rangle = |e;0\rangle$ and (ii) cavity contains one photon and the atom in ground state, i.e., $|\Psi_x^g(0)\rangle = |g;1\rangle$. For these initial conditions, the wave-function (1.47a) evolves according

$$|\Psi_x^e(t)\rangle = \cos(g_\circ t/2)|e;0\rangle + \sin(g_\circ t/2)|g;1\rangle, \tag{1.49a}$$

$$|\Psi_x^g(t)\rangle = \cos(g_\circ t/2)|g;1\rangle - \sin(g_\circ t/2)|e;0\rangle, \tag{1.49b}$$

and describes a coherent exchange of energy between a two-level atom and intracavity field. The frequency of this exchange is given by the the vacuum Rabi splitting $g_\circ$, that is the position-independent part of atom-cavity coupling (1.38). For similar initial conditions, i.e., $|\Psi_y^e(0)\rangle = |e;\bar{0}\rangle$ and $|\Psi_y^g(0)\rangle = |g;\bar{1}\rangle$, the wave-function (1.47b) evolves according

$$|\Psi_y^e(t)\rangle = \cos(g_\circ t/2)|e;\bar{0}\rangle + i\sin(g_\circ t/2)|g;\bar{1}\rangle, \tag{1.50a}$$

$$|\Psi_y^g(t)\rangle = \cos(g_\circ t/2)|g;\bar{1}\rangle + i\sin(g_\circ t/2)|e;\bar{0}\rangle, \tag{1.50b}$$

CHAPTER 1: Interaction of a two-level atom with a single-mode cavity field

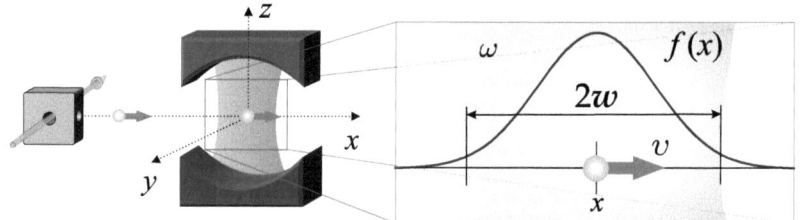

Figure 1.4: Schematic view of the experimental setup. A chain of Rydberg atoms is emitted from the atomic source along the x-axis (in the origin of $y-z$ plane) such that only one atom couples to the cavity at a time. The cavity is characterized by the resonant frequency ω and position-dependent coupling strength $g(x) = g_\circ f(x)$.

where the imaginary factor arises due to orthogonal polarization of the mode C_x with respect to C_y. Note that expressions (1.49) and (1.50) describe the time-varying entanglement between a two-level atomic and cavity photon field. For instance, the Bell state $\frac{1}{\sqrt{2}}(|e;0\rangle + |g;1\rangle)$ is produced from the initial state $|e;0\rangle$ after the interaction time $g_\circ t = \pi/2$ [see (1.49a)]. We conclude, therefore, that by setting an appropriate interaction time and an initial atom-cavity state, we can control the coherent exchange of energy in the coupled atom-cavity system and generate various nonlocal states after the atom is removed from the cavity.

1.2.2 Effective atom-cavity interaction time

So far, we assumed that the atom is at rest inside the cavity such that the atomic center-of-mass position is located at the point $\mathbf{r}'_\circ$ in which the field structure $f(\mathbf{r}'_\circ)$ is maximum. By inspecting the cavity field amplitude (1.27), it is seen that in order to fulfill this requirement the z component must be located in the middle of cavity (at the field antinode), while the components x and y must be located in the origin, i.e., $\mathbf{r}'_\circ = \{0,0,0\}$ [see Fig. 1.2]. By assuming the atom at rest, moreover, we simplified essentially the atom-cavity evolution such that all degrees of freedom associated with the motion of atomic center-of-mass have been excluded.

In practice, however, it is difficult to maintain the atom inside the cavity for a necessary long period of time which, for instance, includes an appropriate atom-cavity interaction time along with atomic trapping and cooling sequences. In the second part of this manuscript we show, moreover, that the atoms utilized in the typical cavity-QED experiments are initialized outside the cavity and emitted with a well controlled velocity by the atomic source as displayed in Fig. 1.4. Without loss of generality, we can assume that an (initialized) atom leaves the atomic source with a constant velocity v along the x-axis such that the atomic trajectory is always in the origin of $z - y$ plane. By this assumption, therefore, the atom probes one single transverse component of cavity field structure while passing through the cavity

$$f(x) = \exp\left[-x^2/w^2\right], \qquad (1.51)$$

which is obtained from the amplitude (1.27) for $z = y = 0$. In this section, we shall analyze how the atom-cavity evolutions (1.49) and (1.50) are affected by an uniform motion of an atom that probes the field amplitude (1.51).

In order to proceed, we consider the Hamiltonian

$$\hat{H}_I^{c.m} = \frac{\hat{p}_x^2}{2\,M} + \hat{H}_I\left(\hat{x}'(t)\right) \equiv \hat{H}^{c.m} + \hat{H}_I\left(\hat{x}'(t)\right), \qquad (1.52)$$

in which the center-of-mass position x' becomes a dynamical variable $\hat{x}'(t)$. The the first term describes the constant kinetic energy associated with atomic center-of-mass motion, and the second term is the atom-cavity interaction Hamiltonian (1.39) with x' being replaced by $\hat{x}'(t)$. According to the Hamiltonian (1.52), moreover, the atomic momentum is influenced only by the second term which describes a coherent exchange of energy between a two-level atom and the cavity field. This exchange implies, in turn, that each process of emission (absorbtion) of a photon, that is characterized by the wave vector $\mathbf{k}$, decreases (increases) the atomic momentum by the amount of $\hbar\mathbf{k}$. In many realistic situation, i.e., when the atom moves with a high velocity, such modifications of atomic momentum are negligible in comparison to the atomic kinetic energy. This physically justified assumption allows us to separate the atomic motion from the Hamiltonian (1.52) and describe it classically. The motion of an atom that moves along the x-axis with a constant velocity v, therefore, is given by the Hamiltonian $H^{c.m} = p_x^2/2\,M$ with $x'(t) = \langle \hat{x}'(t) \rangle$ and $p_x = \langle \hat{p}_x \rangle = M\,\dot{x}'(t)$ which, in turn, yields the equation of motion

$$x'(t) = x'_o + p_x\,t/M = x'_o + v\,t, \qquad (1.53)$$

where $x'_o = \langle \hat{x}'(0) \rangle$ denotes the mean of atomic center-of-mass position at $t = 0$. The obtained equation of motion replaces the dynamical variable $\hat{x}'(t)$ in the Hamiltonian (1.52) and together with function (1.51) leads to the resonant Hamiltonian ($\Delta = 0$)

$$\hat{H}_I^{c.m} = -i\,\hbar\,\frac{g(t)}{2}\left(\hat{\sigma}^\dagger\hat{a} - \hat{\sigma}\,\hat{a}^\dagger\right) - \hbar\,\frac{g(t)}{2}\left(\hat{\sigma}^\dagger\,\hat{b}\,e^{i\,\delta\,t} + \hat{\sigma}\,\hat{b}^\dagger\,e^{-i\,\delta\,t}\right) \qquad (1.54)$$

with the atom-cavity coupling given by

$$g(t) = g_\circ \exp\left[-(x'_o + v\,t)^2/w^2\right]. \qquad (1.55)$$

CHAPTER 1: Interaction of a two-level atom with a single-mode cavity field

As we explained in the previous section, we can safely neglect the second term in the Hamiltonian (1.54) and consider the atom interacting only with the cavity mode C_x

$$\hat{H}_x^{\text{c.m}}(t) = -i\,\hbar\,\frac{g(t)}{2}\left(\hat{\sigma}^\dagger \hat{a} - \hat{\sigma}\,\hat{a}^\dagger\right), \tag{1.56}$$

whenever the condition $\delta \gg g(t)$ is satisfied[5]. For the same condition, moreover, the Hamiltonian (1.52) together with amplitude (1.51) reduces to the resonant interaction of atom with mode C_y

$$\hat{H}_y^{\text{c.m}}(t) = -\hbar\,\frac{g(t)}{2}\left(\hat{\sigma}^\dagger \hat{b} + \hat{\sigma}\,\hat{b}^\dagger\right), \tag{1.57}$$

if the atomic transition is tuned in resonance with this mode ($\Delta = -\delta$). Note that below, we chose the time origin, $t = 0$, when the atom crosses the cavity axis, i.e., $x'(0) = x'_\circ = 0$. With this choose, therefore, the Hamiltonians $\hat{H}_x^{\text{c.m}}(0)$ and $\hat{H}_y^{\text{c.m}}(0)$ coincide with Hamiltonians $\hat{H}_x$ and $\hat{H}_y$, respectively.

In order to solve the Schrödinger equations associated with Hamiltonians (1.56) and (1.57)

$$i\,\hbar\frac{d|\Psi_x(t)\rangle}{dt} = \hat{H}_x^{\text{c.m}}(t)\,|\Psi_x(t)\rangle, \quad i\,\hbar\frac{d|\Psi_y(t)\rangle}{dt} = \hat{H}_y^{\text{c.m}}(t)\,|\Psi_y(t)\rangle, \tag{1.58}$$

we express the wave-functions $|\Psi_x(t)\rangle$ and $|\Psi_y(t)\rangle$ in the limit $t \to \infty$

$$|\Psi_x(\infty)\rangle = \exp\left[-\frac{i}{\hbar}\int_{-\infty}^{+\infty}\hat{H}_x^{\text{c.m}}(t)\,dt\right]|\Psi_x(0)\rangle = \exp\left[-\frac{i}{\hbar}\hat{H}_x^{\text{c.m}}(0)\,t'\right]|\Psi_x(0)\rangle, \tag{1.59a}$$

$$|\Psi_y(\infty)\rangle = \exp\left[-\frac{i}{\hbar}\int_{-\infty}^{+\infty}\hat{H}_y^{\text{c.m}}(t)\,dt\right]|\Psi_y(0)\rangle = \exp\left[-\frac{i}{\hbar}\hat{H}_y^{\text{c.m}}(0)\,t'\right]|\Psi_y(0)\rangle, \tag{1.59b}$$

where

$$t' = \int_{-\infty}^{+\infty}\exp\left[-(v\,t)^2/w^2\right]\,dt = \frac{\sqrt{\pi}\,w}{v}. \tag{1.60}$$

The last parts of expressions (1.59a) and (1.59b) can be interpreted as the formal solution of Schrödinger equations

$$i\,\hbar\frac{d|\Psi_x(t')\rangle}{dt'} = \hat{H}_x^{\text{c.m}}(0)\,|\Psi_x(t')\rangle, \quad i\,\hbar\frac{d|\Psi_y(t')\rangle}{dt'} = \hat{H}_y^{\text{c.m}}(0)\,|\Psi_y(t')\rangle \tag{1.61}$$

associated with (time-independent) Hamiltonians $\hat{H}_x^{\text{c.m}}(0)$ and $\hat{H}_y^{\text{c.m}}(0)$, and which coincide with the Schrödinger equations (1.46) with t being replaced by t'. The ansatz (1.47) applied for the wave-functions $|\Psi_x(t')\rangle$ and $|\Psi_y(t')\rangle$, furthermore, leads to the solutions (1.49) and (1.50) with t being simply replaced by the *effective* atom-cavity interaction time t' defined above. It is clear now, that the atom-cavity evolutions (1.49) and (1.50)

---

[5]In order to show this, one should (i) apply a Galilean boost to the Hamiltonian (1.52) and switch to the frame that moves at the velocity $-v$, (2) expand in series the evolution operator for obtained Hamiltonian and follow the integration procedure we performed in the previous section, and (3) apply a Galilean boost to the obtained Hamiltonian and switch back to the frame that moves at the velocity v.

remain valid for an uniform motion of an atom that probes the field structure (1.51). We can conclude, therefore, that after the atom has crossed the cavity mode, the system has evolved as if the maximum atom-cavity coupling has been applied during the effective interaction time t'. In order to produce the Bell state $\frac{1}{\sqrt{2}}(|e;0\rangle + |g;1\rangle)$ from the initial state $|e;0\rangle$, for instance, an atom should be emitted by the atomic source with the constant velocity $v = 2\,g_\circ\,w/\sqrt{\pi}$ that readily follows from the condition $g_\circ\,t' = \pi/2$.

1.2.3 Damping of Rabi oscillations

The Jaynes-Cummings Hamiltonian (1.44) or (1.45) describes the coupled atom-cavity system in the absence of dissipation, i.e. when the system is well isolated from the environment. In reality, there are two main dissipation processes which should be taken into account: (i) incoherent decay of the excited atomic state caused by its coupling to the free-space electromagnetic background and (ii) cavity field relaxation caused by the losses and scattering on cavity mirrors which lead to the irreversible escape of cavity photons. The realistic atom-cavity evolution, therefore, cannot be described by the Jaynes-Cummings Hamiltonian alone and a more general formalism which handles both coherent and incoherent processes must be considered.

In the presence of dissipations, moreover, the atom-cavity system cannot be treated independently from the environment, which typically has a very broad band spectrum and remains in a thermal equilibrium. However, since we are only interested in the dynamics of coupled atom-cavity system, we trace over all environmental states in order to exclude the degrees of freedom which are not relevant for us. The time evolution of coupled atom-cavity system (at zero temperature) is then described by the master equation [82]

$$\dot{\hat{\rho}} = -\frac{i}{\hbar}\left(\hat{H}_{\text{tot}}\,\hat{\rho} - \hat{\rho}\,\hat{H}_{\text{tot}}^\dagger\right) + \left(\kappa\,\hat{a}\,\hat{\rho}\,\hat{a}^\dagger + \gamma\,\hat{\sigma}\,\hat{\rho}\,\hat{\sigma}^\dagger\right) \qquad (1.62)$$

with

$$\hat{H}_{\text{tot}} = -i\,\hbar\,\frac{g_\circ}{2}\left(\hat{\sigma}^\dagger\hat{a} - \hat{\sigma}\,\hat{a}^\dagger\right) - i\,\hbar\,\frac{1}{2}\left(\gamma\,\hat{\sigma}^\dagger\,\hat{\sigma} + \kappa\,\hat{a}^\dagger\,\hat{a}\right), \qquad (1.63)$$

where γ and κ are the atomic spontaneous emission and cavity relaxation rates, respectively, and $\hat{\rho}$ is the density matrix of system in question. The second parentheses in the master equation is the quantum 'jump' operator, while the second parentheses in the Hamiltonian (1.63) describes the non-hermitian evolution due to the dissipations discussed above and is responsible for the incoherent processes.

In the case of vanishing spontaneous emission and cavity relaxation rates, the solution of master equation (1.62) coincides with Eqs. (1.49) and yields the oscillatory population probability

$$P_e(t) = |\langle e|\Psi_x^e(t)\rangle|^2 = \frac{1 + \cos(g_\circ\,t)}{2} \qquad (1.64)$$

CHAPTER 1: Interaction of a two-level atom with a single-mode cavity field

for initially excited atom and empty cavity. Using the master equation, we like to calculate how the amplitude of these oscillations modifies in the case of non-vanishing spontaneous emission and cavity relaxation rates. Similarly to the section 1.2.1, we introduce an ansatz for the wave-function that describes the evolution of atom-cavity system and substitute it into the master equation. Obviously, there are only three relevant states which are involved in this evolution, namely, two 'one-quanta' states $|e; 0\rangle$, $|g; 1\rangle$, and one 'null-quanta' state $|g; 0\rangle$. Since the coupling between these two manifolds irreversibly populates the zero-quanta subspace and thus is not interesting for us, we can assume the *unnormalized* one-quantum ansatz

$$|\Psi_u(t)\rangle = \Psi_u^{e,0}(t)|e; 0\rangle + \Psi_u^{g,1}(t)|g; 1\rangle \quad \text{with} \quad |\Psi_u(0)\rangle = |e; 0\rangle . \tag{1.65}$$

By substituting the density operator $\hat{\rho} = |\Psi_u(t)\rangle\langle\Psi_u(t)|$ into the master equation, we find that the evolution of $|\Psi_u(t)\rangle$ is governed by the effective non-hermitian Schrödinger equation

$$i\hbar \frac{d|\Psi_u(t)\rangle}{dt} = \hat{H}_{\text{tot}} |\Psi_u(t)\rangle , \tag{1.66}$$

which implies the system of differential equations for amplitudes $\Psi_u^{e,0}(t)$ and $\Psi_u^{g,1}(t)$

$$\dot{\Psi}_u^{e,0}(t) = -\frac{g_\circ}{2} \Psi_u^{g,1} - \frac{\gamma}{2} \Psi_u^{e,0}(t), \quad \dot{\Psi}_u^{g,1}(t) = \frac{g_\circ}{2} \Psi_u^{e,0} - \frac{\kappa}{2} \Psi_u^{g,1}(t), \tag{1.67}$$

where dot denotes the time derivative and where $\Psi_u^{e,0}(0) = 1$ and $\Psi_u^{g,1}(0) = 0$ are the initial conditions. At this point, it is useful to distinguish between two qualitatively different regimes. In the first one, the decay rates κ and γ dominate over the atom-cavity coupling frequency $g_\circ$. This is conventionally called the weak coupling regime (or bad cavity regime). In contrast, the strong coupling regime (or good cavity regime) is characterized by the fact that the coherent interaction between the atom and the cavity mode dominates over the irreversible dissipation processes, i.e., $g_\circ \gg \kappa$ and $g_\circ \gg \gamma$.

Integrating the second equation from Eqs. (1.67), we obtain the formal solution

$$\Psi_u^{g,1}(t) = \frac{g_\circ}{2} \int_0^t \Psi_u^{e,0}(t') \exp\left[\frac{k}{2}(t'-t)\right] dt' . \tag{1.68}$$

The Eqs. (1.67), moreover, imply that $\Psi_u^{e,0}(t)$ is slow-varying under the weak coupling conditions, i.e., $g_\circ \ll \kappa$ and $g_\circ \ll \gamma$. Therefore, we can evaluate $\Psi_u^{e,0}(t)$ at the time $t' = t$ and remove it from the above integral. For $t \gg \kappa^{-1}$, the remaining integral yields $\Psi_u^{g,1}(t) \approx g_\circ \Psi_u^{e,0}(t)/\kappa$, which being substituted into the first of Eqs. (1.67) and integrated, implies

$$\Psi_u^{e,0}(t) = \exp\left[-\frac{1}{2}\left(\gamma + \frac{g_\circ^2}{\kappa}\right)t\right], \quad \text{and} \quad P_e^{\text{weak}}(t) = \exp\left[-\left(\gamma + \frac{g_\circ^2}{\kappa}\right)t\right] . \tag{1.69}$$

According to the last expression, the upper electronic state population undergoes an exponential decay at the rate $\gamma + g_\circ^2/\kappa$ [compare to (1.64)], where the cavity contribution is $g_\circ^2/\kappa$. For a sufficiently high quality factor of cavity mirrors, this contribution predicts a considerable enhancement of spontaneous emission rate as compared to its free space value [10].

Below, we consider the strong atom-cavity coupling regime, i.e., when a photon emitted into the cavity mode is likely to be reabsorbed before it escapes from the resonator. The general solution of Eqs. (1.67) for arbitrary $g_\circ$, κ, and γ, is of the form

$$\Psi_u^{e,0}(t) = C_+ X_+ e^{\lambda_+ t} + C_- X_- e^{\lambda_- t}, \quad \Psi_u^{g,1}(t) = C_+ Y_+ e^{\lambda_+ t} + C_- Y_- e^{\lambda_- t}, \qquad (1.70)$$

where $\lambda_\pm$ and vectors $\{X, Y\}_\pm$ are the respective eigenvalues and eigenvectors of the matrix $\begin{pmatrix} -\gamma/2 & -g_\circ/2 \\ g_\circ/2 & -\kappa/2 \end{pmatrix}$, and where the constants $C_\pm$ are determined by the initial conditions. By evaluating the eigenvalues of above matrix, we obtain the expressions

$$\lambda_\pm = \frac{1}{4}\left(-\kappa - \gamma \pm \sqrt{(\gamma-\kappa)^2 - 4 g_\circ^2}\right) \qquad (1.71)$$

which for the strong coupling conditions reduce to $\lambda_\pm = \frac{1}{4}(-\kappa - \gamma \pm 2 i\, g_\circ)$. Owning to this result, we can conclude that the amplitude of imaginary part of exponents from Eqs. (1.70) is much larger than that of the real part. This implies, in turn, that the evolution of upper state population will consist of oscillations at the vacuum Rabi frequency which slowly decay in time.

Furthermore, by evaluating the eigenvectors of above matrix and taking into account the initial conditions $\Psi_u^{e,0}(0) = 1$ and $\Psi_u^{g,1}(0) = 0$, we obtain the evolution of amplitude $\Psi_u^{e,0}(t)$

$$\Psi_u^{e,0}(t) = \frac{e^{-\frac{1}{4}(\kappa+\gamma)t}}{4 i\, g_\circ}\left((\kappa - \gamma + 2 i\, g_\circ) e^{i\, g_\circ t/2} - (\kappa - \gamma - 2 i\, g_\circ) e^{-i\, g_\circ t/2}\right) \qquad (1.72)$$

and which implies, in turn, the population probability of upper electronic state

$$P_e^{\text{strong}}(t) = |\langle e|\Psi_u(t)\rangle|^2 = e^{-\frac{1}{2}(\kappa+\gamma)t}\left(\frac{1+\cos(g_\circ t)}{2}\right). \qquad (1.73)$$

Indeed, the evolution of upper state population consists of oscillations at the vacuum Rabi frequency which exponentially decay in time. By comparing the probability (1.73) that we obtained from the master equation with the the idealized Jaynes-Cummings probability (1.64), it is clearly seen how the amplitude of these oscillations modifies in the case of non-vanishing spontaneous emission and cavity relaxation rates. We conclude, therefore, that the main modification is introduced due to the slow exponential decay given by the sum of spontaneous emission and cavity relaxation rates.

1.3 Summary

In this chapter, we introduced and described the interaction of a two-level atom with a single-mode monochromatic light in the quantum regime. First, we obtained the

CHAPTER 1: Interaction of a two-level atom with a single-mode cavity field

expressions (1.13) and (1.15) for energy and electromagnetic field which both evolve inside a planar or spherical cavity. We demonstrated, moreover, that the light field of a spherical cavity supports two transversal components encoded in the cavity field structure (1.27). Next, we derived the Hamiltonian (1.39) that governs the interaction of confined photon field coupled to one two-level atom. We explicitly assumed that the cavity supports two linearly and orthogonally polarized modes of light (C_x and C_y) separated in frequencies by the birefringent splitting δ, while the atom emits or absorbs the circularly polarized light during its transition.

We analyzed, furthermore, the situation in which the atomic transition frequency matches the frequency of one of cavity modes and we found that the atom-cavity evolution is governed by the Jaynes-Cummings Hamiltonian (1.44) or (1.45), whenever the birefringent splitting is sufficiently detuned with respect to the Rabi vacuum splitting. We found out that $A - C_x$ and $A - C_y$ evolutions are given by the expressions

$$|e;0\rangle \to \cos(g_\circ t/2)|e;0\rangle + \sin(g_\circ t/2)|g;1\rangle, \tag{1.74a}$$

$$|g;1\rangle \to \cos(g_\circ t/2)|g;1\rangle - \sin(g_\circ t/2)|e;0\rangle, \tag{1.74b}$$

$$|e;\bar{0}\rangle \to \cos(g_\circ t/2)|e;\bar{0}\rangle + i\sin(g_\circ t/2)|g;\bar{1}\rangle, \tag{1.75a}$$

$$|g;\bar{1}\rangle \to \cos(g_\circ t/2)|g;\bar{1}\rangle + i\sin(g_\circ t/2)|e;\bar{0}\rangle, \tag{1.75b}$$

respectively, and both evolutions describe the time-varying entanglement of a two-level atom with cavity photon field.

We analyzed then the situation in which an atom is prepared and initialized outside the cavity and emitted with a well controlled velocity such that its trajectory crosses the cavity at the antinode. In this situation, the atom probes one single transverse component of intracavity field while passing through the cavity. We found that the atom-cavity evolution for an uniformly moving atoms is still given by the expressions (1.74) and (1.75), however, the interaction time t must be replaced by the effective interaction time t' given by (1.60). Finally, we analyzed the effects of spontaneous atomic emission and cavity relaxation in order to understand how the energy exchange of coupled atom-cavity system evolves in realistic environments.

In the next chapters, we shall interpret the atoms and cavity photons as qubits and from this perspective, the atom-cavity evolution will enable us to generate various multipartite entangled states with atomic qubits which pass sequentially through the cavity.

Chapter 2

Interaction of three-level Λ-type atoms with cavity and laser fields

In the previous chapter, we described the interaction of a single atom with monochromatic light field in the quantum regime. We assumed the validity of two-level approximation and derived the Jaynes-Cummings Hamiltonian that governs the evolution of coupled atom-light system. By considering the resonant atom-cavity interaction regime, furthermore, we found that this evolution provides a controllable entanglement mechanism of a two-level atom with the cavity photon field and which, therefore, offers an excellent framework for quantum information processing (see the third part).

In section 1.2.1, moreover, we showed that if one of cavity modes is tuned in resonance with the atomic transition, then the second cavity mode produces only the phase shifts (1.43) which can be neglected whenever the condition $\delta \gg g_\circ$ is satisfied. This observation allowed us to exclude the contribution of term quadratic in $g_\circ$ and describe the atom-cavity evolution by using only the resonant Hamiltonian (1.44) or (1.45). Instead of tuning one of the cavity modes in resonance with the atom, we could slightly detune both modes from the resonance such that the effective Hamiltonian (1.41) contains only the terms quadratic in $g_\circ$. This situation, in which there are no linear in $g_\circ$ terms, looks to be useless from the first sight since no energy exchange between atom and cavity occurs. However, if we place two or more atoms in the (detuned) cavity such that all of them are simultaneously coupled to the same light modes, then the dipole-dipole interaction can be realized as a consequence of the virtual cavity photon exchange between the atoms. This (cavity mediated) energy exchange, moreover, is less sensitive with regard to decoherence effects since the cavity remain almost unpopulated during the entire evolution. The scheme for generation of atom-atom entanglement that is based on the described situation has been proposed in Ref. [21] and, later on, realized experimentally in the framework of microwave cavity-QED [39].

CHAPTER 2: Interaction of three-level Λ-type atoms with cavity and laser fields

In this chapter, we first describe the interaction of a single atom with monochromatic light field in the semiclassical regime. Right afterwards, we introduce and explain our scheme for generation of atomic multipartite entangled states which is based on the off-resonant interaction regime of three-level atoms placed inside the cavity and coupled simultaneously to a laser beam [62, 63]. We perform the adiabatic elimination and find that the evolution of three-level atoms is reduced to the evolution of effectively two-level atoms which interact with each other via a two-photon exchange. Finally, by assuming that the atoms are moving uniformly through the cavity and are equally distanced with respect to each other, we determine how this interaction is characterized by atomic velocities and inter-atomic distances.

2.1 Semiclassical atom-field interaction

The Hamiltonians (1.44) and (1.45) which we introduced in the previous chapter, describe quantum mechanically both the internal atomic structure and the intracavity light field. In many situations, however, it is justified to treat the light field classically. For instance, a laser beam is a coherent radiation that consists of a large number of photons such that individual photon fluctuations become negligible. Therefore, to a very good approximation which is known as *semiclassical* approximation, the electromagnetic field produced by such a laser beam is described classically by a plane (monochromatic) light wave that satisfies the Maxwell equations (1.1), while the internal atomic structure is described quantum mechanically by the Hamiltonian (1.32).

In the previous chapter, moreover, we analyzed the situation in which the electromagnetic field propagates between the (planar or spherical) mirrors of a cavity. In contrast to the situation with a cavity, the laser beam is well described by a plane (monochromatic) wave which propagates infinitely in one spatial direction. Without loss of generality, we can assume that the direction of propagation coincides with the positive direction of y-axis and, therefore, the laser beam characterized by the frequency ω_L is described by the electromagnetic field

$$\mathbf{E}^s(\mathbf{r},t) = i\, \mathcal{E}_\circ \left(\boldsymbol{\epsilon}_L\, e^{i\,(k\,y - \omega_L t)} - \boldsymbol{\epsilon}_L^*\, e^{-i\,(k\,y - \omega_L t)} \right), \tag{2.1}$$

which satisfies the equation (1.2) and where $k = \omega_L/c$ denotes the wave number, $\boldsymbol{\epsilon}_L$ is the polarization, and $\mathcal{E}_\circ$ is the real amplitude of laser field. Similarly to the previous chapter, we assume that the atom consists of a hard-core and one valence electron and it interacts with the field produced by a laser beam via the dipole-field Hamiltonian[1]

$$\hat{H}^s_{\text{int}} = -\mathrm{q}\,\hat{\mathbf{r}}\cdot \mathbf{E}^s(\mathbf{r}',t), \tag{2.2}$$

---
[1]Since the wavelength of laser light is much larger than the extent of wave-function associated to the atom, the dipole approximation remains valid here as well.

where $\mathbf{r}' = \{0, y', 0\}$ denotes the atomic center-of-mass position at which the atom is localized during the interaction.

With the help of orthogonality and completeness relations (1.31), we express the interaction Hamiltonian $\hat{H}^s_{\text{a-f}} = \hat{H}_a + \hat{H}^s_{\text{int}}$ in the form

$$\hat{H}^s_{\text{a-f}} = \hat{H}_a - i\, d\, \mathcal{E}_\circ \left(\hat{\sigma}^\dagger\, \boldsymbol{\epsilon}_a^* + \hat{\sigma}\, \boldsymbol{\epsilon}_a \right) \cdot \left(\boldsymbol{\epsilon}_L\, e^{i\,(k\,y' - \omega_L t)} - \boldsymbol{\epsilon}_L^*\, e^{-i\,(k\,y' - \omega_L t)} \right), \qquad (2.3)$$

where $\hat{\sigma} = |g\rangle\langle e|$ and $\hat{\sigma}^\dagger = |e\rangle\langle g|$ denote the excitation lowering and rising operators, respectively, and which is further simplified by switching to the interaction picture with respect to the time-independent part $\hat{H}_a$

$$\hat{H}^s_I = \hat{U}_a^\dagger \left(\hat{H}^s_{\text{a-f}} - \hat{H}_a \right) \hat{U}_a = -i\hbar\, \frac{\Omega_\circ}{2} \left(\hat{\sigma}^\dagger\, e^{i\,\Delta_L t}\, e^{-i\varphi} - \hat{\sigma}\, e^{-i\,\Delta_L t}\, e^{i\varphi} \right). \qquad (2.4)$$

In this Hamiltonian, $\Delta_L = \omega_a - \omega_L$ is the difference between the atomic transition frequency and the laser field frequency, and the Rabi frequency (atom-laser coupling)

$$\Omega_\circ = \frac{2}{\hbar}\, d\, \mathcal{E}_\circ\, |\boldsymbol{\epsilon}_a^* \cdot \boldsymbol{\epsilon}_L| \qquad (2.5)$$

has been also introduced. Note that in the Hamiltonian (2.4), we have chosen the splitting $y' = -(\varphi + \vartheta)/k$ such that $e^{i\vartheta}$ coincides with the phase of $\boldsymbol{\epsilon}_a^* \cdot \boldsymbol{\epsilon}_L$, and moreover, we neglected the terms $\hat{\sigma}^\dagger\, e^{-i(k\,y' - [\omega_L + \omega_a]t)}$ and $\hat{\sigma}\, e^{i(k\,y' - [\omega_L + \omega_a]t)}$ because of the rotating wave approximation.

Hamiltonian (2.4) governs the interaction of a two-level atom with the field produced by a laser beam. The physically interesting regime of atom-laser evolution is the resonant interaction regime, in which the atom-laser detuning Δ_L vanishes and the Hamiltonian (2.4) becomes

$$\hat{H}^s_I = -i\hbar\, \frac{\Omega_\circ}{2} \left(\hat{\sigma}^\dagger\, e^{-i\varphi} - \hat{\sigma}\, e^{i\varphi} \right). \qquad (2.6)$$

In order to determine the behavior of atomic population, we shall calculate the evolution operator $\hat{U}^s_I(t) = \exp\left[-\frac{i}{\hbar} \hat{H}^s_I t \right]$ associated with the Hamiltonian (2.6). Using the vector representation $|e\rangle = \begin{pmatrix} 1 \\ 0 \end{pmatrix}$ and $|g\rangle = \begin{pmatrix} 0 \\ 1 \end{pmatrix}$, it can be shown that the evolution operator associated with the Hamiltonian (2.6) is given by

$$\hat{U}^s_I(t) = \begin{pmatrix} 1 & 0 \\ 0 & 1 \end{pmatrix} \cos\left(\frac{\Omega_\circ}{2} t \right) + \begin{pmatrix} 0 & -e^{-i\varphi} \\ e^{i\varphi} & 0 \end{pmatrix} \sin\left(\frac{\Omega_\circ}{2} t \right). \qquad (2.7)$$

The matrix (2.7) together with the initial atomic state $|\Phi(0)\rangle$ determine the time-evolution associated with the Hamiltonian (2.6). We are interested, however, in the atom-laser evolution obtained for two initial states: (i) atom is excited, i.e., $|\Phi_e(0)\rangle = |e\rangle$

CHAPTER 2: Interaction of three-level Λ-type atoms with cavity and laser fields

and (ii) atom in the ground state, i.e., $|\Phi_g(0)\rangle = |g\rangle$. For these initial conditions, the expression (2.7) implies the time-evolution

$$|\Phi_e(t)\rangle = \cos(\Omega_\circ t/2)|e\rangle + \sin(\Omega_\circ t/2)\, e^{i\varphi}|g\rangle, \qquad (2.8a)$$

$$|\Phi_g(t)\rangle = \cos(\Omega_\circ t/2)|g\rangle - \sin(\Omega_\circ t/2)\, e^{-i\varphi}|e\rangle, \qquad (2.8b)$$

and describes the oscillations of electronic population between its ground and excited states such that the frequency of this exchange is given by the Rabi frequency (2.5). For instance, the superposition $\frac{1}{\sqrt{2}}\left(|e\rangle + e^{i\varphi}|g\rangle\right)$ is produced from the initial state $|e\rangle$ after the interaction time $\Omega_\circ t = \pi/2$ [see (2.8a)]. We conclude, therefore, that by setting an appropriate interaction time and an initial atomic state, we can generate any atomic superposition.

We assumed that the atom is at rest and the atomic center-of-mass position is located at the point $\mathbf{r}' = \{0, y', 0\}$ in which the splitting $y' = -(\phi + \vartheta)/k$ is realized [see (2.4)]. In practice, however, it is difficult to maintain the atom in a fixed position for the entire atom-laser interaction time. In the second part of this manuscript we show, moreover, that the atoms utilized in the typical cavity-QED experiments are initialized and emitted with a well controlled velocity by the atomic source. Without loss of generality, we can consider an (initialized) atom that leaves the atomic source with a constant velocity v along the x-axis such that atomic trajectory is determined by $\mathbf{r}'(t) = \{v\,t, y', 0\}$. By following the arguments of section 1.1.1, however, it can be shown that the field structure of a laser beam supports two transverse components

$$\mathcal{E}(\mathbf{r}) = \mathcal{E}_\circ \exp\left[-(x^2 + z^2)/w_L^2\right], \qquad (2.9)$$

where w_L denotes the waist of laser beam. This position-dependent amplitude, in turn, implies the position-dependent Rabi frequency

$$\Omega(x) = \frac{2}{\hbar}\,\mathbf{d}\,\mathcal{E}(\mathbf{r})\,|\boldsymbol{\epsilon}_a^* \cdot \boldsymbol{\epsilon}_L| = \Omega_\circ \exp\left[-x^2/w_L^2\right], \qquad (2.10)$$

obtained for $z = 0$ and which in the case of uniform atomic motion along the x-axis, becomes

$$\Omega(t) = \Omega_\circ \exp\left[-(v\,t)^2/w_L^2\right]. \qquad (2.11)$$

By performing the same analysis as in section 1.2.2, it can be shown that the time-evolution of an uniformly moving atom coincides with (2.8) with t being replaced by t' defined as

$$t' = \int_{-\infty}^{+\infty} \exp\left[-(v\,t)^2/w_L^2\right]\,dt = \frac{\sqrt{\pi}\,w_L}{v}. \qquad (2.12)$$

After the atom has crossed the laser beam, therefore, the atomic state has evolved as if the maximum atom-laser coupling $\Omega_\circ$ has been applied during the effective interaction

30

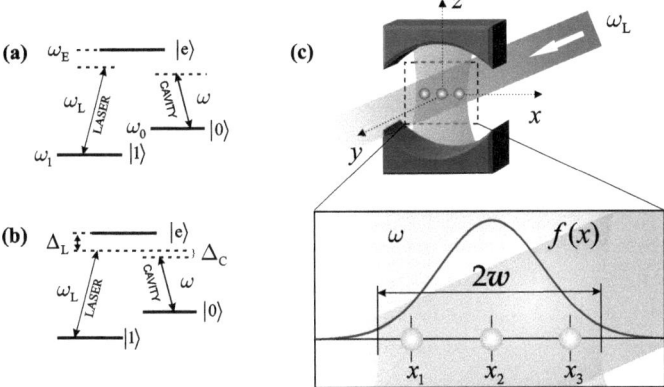

Figure 2.1: (a) The atomic three-level Λ-type configuration in the Schrödinger picture and (b) in the interaction picture. (c) Schematic view of experimental setup. A chain of neutral atoms is localized in the cavity along the x-axis (in the origin of $y-z$ plane). The cavity is characterized by the resonant frequency ω and position-dependent coupling strength $g(x)$, while the laser is characterized by the frequency ω_L and position-dependent coupling $\Omega(x)$.

time t'. In order to produce the superposition $\frac{1}{\sqrt{2}}\left(|e\rangle + e^{i\varphi}|g\rangle\right)$, for example, an atom should be emitted by the atomic source with the constant velocity $v = 2\,\Omega_\circ\,w_L/\sqrt{\pi}$, which readily follows from the condition $\Omega_\circ\,t' = \pi/2$.

Finally, we like to recall that the Hamiltonian (2.4) is expressed in the interaction picture, in which the atomic Hamiltonian $\hat{H}_a$ does not enter explicitly. In many situations, however, it is convenient to consider the picture in which Hamiltonian depends explicitly on the atomic frequencies ω_e, ω_g and laser frequency ω_L. In order to switch to such a picture, we act on (2.4) from left and right sides with operators $\hat{U}_a$ and $\hat{U}_a^\dagger$, respectively

$$\hat{H}^s = \hat{H}_a + \hat{U}_a\,\hat{H}_I^s\,\hat{U}_a^\dagger = \hbar\left[\omega_g\,|g\rangle\langle g| + \omega_e\,|e\rangle\langle e| - i\,\frac{\Omega_\circ}{2}\left(\hat{\sigma}^\dagger\,e^{-i\omega_L t}\,e^{-i\varphi} - \hat{\sigma}\,e^{i\omega_L t}\,e^{i\varphi}\right)\right]. \quad (2.13)$$

In contrast to the Hamiltonian (2.4) which depends solely on the detuning Δ_L, the obtained Hamiltonian depends on the atomic frequencies ω_e, ω_g and laser frequency ω_L.

2.2 Generation of multipartite entangled states

It is clear that any particular scheme to encode one single qubit in the level structure of an atom, depends crucially on the frequency that is supported by the cavity that is used to mediate the interaction between two or more atoms. For an optical cavity, the two levels which encode a qubit are usually selected to be the ground and first excited states of an atom since they are (usually) separated by an optical transition frequency ($\sim 1~\mu$m). In contrast to the stable ground state, however, the first excited state decays very fast ($\sim 10^{-9}$ s) and is not useful for encoding a qubit. For this reason, therefore, three-level Λ-type atoms are widely utilized in the experiments in which optical cavities are used. A three-level atom in the Λ-type configuration is displayed in Fig. 2.1(a) and it allows to encode one single qubit state in the long-living metastable state $|0\rangle$ and the stable ground state $|1\rangle$, respectively. In such an atom, moreover, the transition $|0\rangle \leftrightarrow |1\rangle$ is (electric-dipole) forbidden due to the angular momentum and parity selection rules.

By this encoding scheme, moreover, the cavity is coupled to the optical $|0\rangle \leftrightarrow |e\rangle$ transition and an intermediate excitation coupled to the $|e\rangle \leftrightarrow |1\rangle$ transition is further needed to link the states $|0\rangle$ and $|1\rangle$ via a two-photon exchange process. For this purpose, a laser beam is coupled to the transition $|e\rangle \leftrightarrow |1\rangle$ as displayed in Fig. 2.1(a) and is adjusted together with the cavity parameters such that the (fast decaying) excited state $|e\rangle$ remains almost unpopulated during the combined atom-cavity-laser interaction (see below). In Fig. 2.1(c) we display a schematic view of experimental setup that realizes the interaction between two or more qubits being encoded in three-level atoms according to our scheme [62, 63]. This setup consists of a cavity field that acts along the z-axis, a laser beam that acts along the y-axis, and a chain of distanced atoms which is localized along the x-axis. Moreover, the cavity is characterized by the resonant frequency ω and position-dependent coupling [see (1.51)]

$$g(x) = g_\circ f(x) = g_\circ \exp\left[-x^2/w^2\right], \tag{2.14}$$

while the laser is characterized by the frequency ω_L and position-dependent coupling (2.10). Without going much into details, in this section we discuss the basic idea to generate multipartite entangled states within a chain of three-level atoms coupled simultaneously to the cavity and laser fields.

Before we turn to the combined cavity and laser mediated interaction between N three-level atoms, it is useful to explain the cavity-mediated interaction between N two-level atoms corresponding to the states $|0\rangle$ and $|e\rangle$ in our scheme, which are prepared initially in the product state $|e_1, 0_2, \ldots, 0_N\rangle$ and where the numbering corresponds to the (increasing) coordinates $x_1, x_2, \ldots, x_N$ of atoms along the x-axis [see Fig. 2.1(c)]. For simplicity, however, let us first consider a chain of two such atoms prepared in the product state $|e_1, 0_2\rangle$. In this case, both atoms interact due to the cavity-induced exchange of a photon according to the evolution sequence

$$|e_1, 0_2; n\rangle \begin{array}{c} \nearrow |0_1, 0_2; n+1\rangle \searrow \\ \searrow |e_1, e_2; n-1\rangle \nearrow \end{array} |0_1, e_2; n\rangle, \tag{2.15}$$

if there were n photons initially in the cavity mode. This sequence describes a process in which one photon is emitted by the first and absorbed by the second atom, so that the final atomic state is independent of the number of initial cavity photons. For an initially empty cavity $n = 0$, therefore, the above sequence reduces to

$$|e_1, 0_2; 0\rangle \to |0_1, 0_2; 1\rangle \to |0_1, e_2; 0\rangle, \quad (2.16)$$

which can be expressed in the short form $|e_1, 0_2\rangle \to |0_1, e_2\rangle$ and interpreted as the (cavity-mediated) dipole-dipole interaction between two atoms.

By following similar line of reasoning, the initial atomic state $|e_1, 0_2, \ldots, 0_N\rangle$ of N atoms evolves according to the sequence

$$
\begin{aligned}
|e_1, 0_2, \ldots, 0_N; 0\rangle \to |0_1, \ldots, 0_N; 1\rangle &\to |0_1, e_2, 0_3, \ldots, 0_N; 0\rangle \\
&\searrow |0_1, 0_2, e_3, \ldots, 0_N; 0\rangle \\
&\quad \vdots \\
&\quad |0_1, 0_2, 0_3, \ldots, e_N; 0\rangle,
\end{aligned} \quad (2.17)
$$

such that each of final states $|0_1, e_2, \ldots, 0_N\rangle$, ..., $|0_1, 0_2, \ldots, e_N\rangle$ can have the same probability to occur if the atom-cavity interaction time is set properly. As in the case of sequence (2.16), the atoms in the chain interact due to cavity-induced exchange of a single photon between the initially excited atom and one of the $N-1$ other atoms. This photon exchange, moreover, requires a rather large detuning between the atomic $|0\rangle \leftrightarrow |e\rangle$ transition and the resonant frequency of cavity mode

$$|(\omega_E - \omega_0) - \omega| \gg g(x_i), \quad i = 1, \ldots, N \quad (2.18)$$

such that the cavity remains almost unpopulated in the course of interaction and the intermediate state $|0_1, \ldots, 0_N; 1\rangle$ becomes virtual [21].

Recall that according to our qubit encoding scheme, in which the excited state $|e\rangle$ is not a part of qubit, the atomic chain is prepared in the product state $|1_1, 0_2, \ldots, 0_N\rangle$. In order to realize the cavity-mediated interaction as explained above, the detuned laser beam coupled to the $|1\rangle \leftrightarrow |e\rangle$ transition is utilized and the entire chain of three-level atoms becomes coupled simultaneously to the both cavity and laser fields. In this case, the initial atomic state $|1_1, 0_2, \ldots, 0_N\rangle$ evolves according to the sequence of intermediate states $|1_1, 0_2, \ldots, 0_N; 0\rangle \to |e_1, 0_2, \ldots, 0_N; 0\rangle \to$

$$
\begin{aligned}
\to |0_1, 0_2, \ldots, 0_N; 1\rangle &\to |0_1, e_2, 0_3 \ldots, 0_N; 0\rangle \to |0_1, 1_2, 0_3, \ldots, 0_N; 0\rangle \\
&\searrow |0_1, 0_2, e_3, \ldots, 0_N; 0\rangle \to |0_1, 0_2, 1_3, \ldots, 0_N; 0\rangle \\
&\quad \vdots \qquad\qquad\qquad\qquad \vdots \\
&\quad |0_1, 0_2, 0_3, \ldots, e_N; 0\rangle \to |0_1, 0_2, 0_3, \ldots, 1_N; 0\rangle
\end{aligned} \quad (2.19)
$$

into the one of final states $|0_1, 1_2, \ldots, 0_N\rangle, \ldots, |0_1, 0_2, \ldots, 1_N\rangle$, which can have the same probability to occur if the atom-cavity-laser interaction time is set properly. Recall that in the case of sequence (2.17), we imposed the condition (2.18) in order to treat the intermediate state $|0_1, \ldots, 0_N; 1\rangle$ as being (almost) unpopulated and simplify the sequence. In the case of sequence (2.19), similarly, it is necessary to impose the condition

$$|(\omega_E - \omega_1) - \omega_L| \gg \Omega(x_i), \quad i = 1, \ldots, N \qquad (2.20)$$

for the detuning between the atomic $|1\rangle \leftrightarrow |e\rangle$ transition and laser frequencies, which ensures that the states $|e_1, 0_2, \ldots, 0_N\rangle, \ldots, |0_1, 0_2, \ldots, e_N\rangle$ remain (almost) unpopulated during the atom-laser interaction. In the following, we shall omit all these unpopulated (intermediate) states along with the factored cavity state $|0\rangle$ and express the sequence (2.19) in the short form

$$\begin{aligned} |1_1, 0_2, \ldots, 0_N\rangle &\to |0_1, 1_2, 0_3, \ldots, 0_N\rangle \\ &\searrow |0_1, 0_2, 1_3, \ldots, 0_N\rangle \\ &\quad \vdots \\ &\quad |0_1, 0_2, 0_3, \ldots, 1_N\rangle. \end{aligned} \qquad (2.21)$$

We conclude, therefore, that the evolution of atomic state $|1_1, 0_2, \ldots, 0_N\rangle$ subjected to the detuned cavity and laser fields is described by the sequence (2.21) which is characterized by the composite wave-function

$$|\Phi_N(t)\rangle = C_1(t)|1_1, 0_2, \ldots, 0_N\rangle + C_2(t)|0_1, 1_2, \ldots, 0_N\rangle + \ldots + C_N(t)|0_1, 0_2, \ldots, 1_N\rangle \qquad (2.22)$$

with $C_1(t), \ldots, C_N(t)$ being the complex amplitudes such that $\sum_{i=1}^{N} |C_i(t)|^2 = 1$, and where t denotes the interaction time of atomic chain with cavity-laser fields. For the interaction period $t = \tau$ such that $|C_i(\tau)| = 1/\sqrt{N}$, moreover, the wave-function (2.22) reduces to the (so-called) W state [22]

$$|\Psi_N^W\rangle = \frac{1}{\sqrt{N}}(e^{i\phi}\overbrace{|1_1, 0_2, \ldots, 0_N\rangle + |0_1, 1_2, \ldots, 0_N\rangle + \ldots + |0_1, 0_2, \ldots, 1_N\rangle}^{N \text{ terms}}), \qquad (2.23)$$

which is relevant for such practical applications like quantum dense coding and quantum key distribution [4].

In this section, we assumed that the atoms are localized in the cavity along the x-axis (in the origin of $y - z$ plane) such that the atomic center-of-mass positions correspond to the (increasing) coordinates $x_1, x_2, \ldots, x_N$ [see Fig. 2.1(c)]. In practice, however, it is difficult to maintain the atoms at fixed positions for the entire atom-cavity-laser interaction time. In the second part of this manuscript we show, moreover, that the atoms utilized in the typical cavity-QED experiments are initialized outside the cavity and transported through the cavity by means of an optical lattice (conveyor belt, see

2.3. Combined atom-cavity-laser interaction

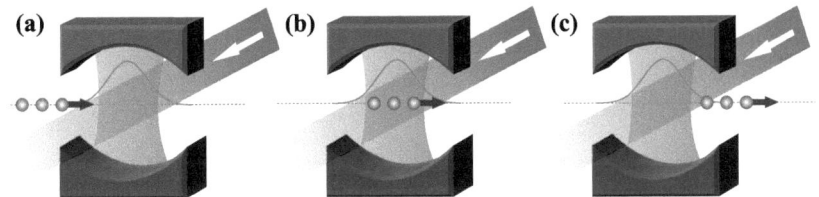

Figure 2.2: Three snapshots of the sequence that generates the W-class state with a chain of three atoms. See the text for explanations.

section 4.2.1). Below, we shall consider an initialized chain of equally distanced atoms that is transported through the cavity with a constant velocity v along the x-axis such that the qubits are loaded to the cavity in the reverse order, i.e. qubit with subscript '1' corresponds to the last atom inside the chain. This procedure is displayed in Fig. 2.2 as a sequence of three snapshots which indicate the atoms (a) outside the cavity (before interaction), (b) inside the cavity (during the interaction), and (c) far away from the cavity (no interaction).

In the next section, we shall analyze in details how the cavity-laser mediated evolution (2.21) depends on the velocity v and the inter-atomic distance $d = x_{i+1}^o - x_i^o$ when the atomic chain is conveyed through the cavity. In order to take into account these two parameters, we shall consider the position-dependent couplings (2.10) and (2.14), which give rise to the time-dependent coupling between the i-th atom and the cavity field ($i = 1, \ldots, N$)

$$g_i(t) = g(x_i^o + v\,t) = g_\circ \exp\left[-(x_i^o + v\,t)^2/w^2\right], \tag{2.24}$$

and to the time-dependent coupling between the i-th atom and the laser field

$$\Omega_i(t) = \Omega(x_i^o + v\,t) = \Omega_\circ \exp\left[-(x_i^o + v\,t)^2/w_L^2\right], \tag{2.25}$$

if the atom moves uniformly and where x_i^o denotes the initial position of i-th atom.

2.3 Combined atom-cavity-laser interaction

While sequence (2.19) displays the basic concept of how the cavity-laser mediated interaction is achieved between the atoms, we still have to analyze this coupling quantitatively as to understand how to control it in practice. For this purpose, we shall adiabatically eliminate the intermediate states $|0_1, 0_2, \ldots, 0_N; 1\rangle$ and $|e_1, 0_2, \ldots, 0_N; 0\rangle$, $\ldots$, $|0_1, 0_2, \ldots, e_N; 0\rangle$ from the sequence (2.19). This shall lead to an effective Hamiltonian that governs the time evolution of N atoms which interact with each other according

CHAPTER 2: Interaction of three-level Λ-type atoms with cavity and laser fields

to the simplified sequence (2.21). To proceed, let us introduce the short-hand notation

$$
\begin{aligned}
|\mathbf{V}_1\rangle \to |\mathbf{V}_{N+1}\rangle &\to |\mathbf{V}_0\rangle \to |\mathbf{V}_{N+2}\rangle \to |\mathbf{V}_2\rangle \\
&\searrow |\mathbf{V}_{N+3}\rangle \to |\mathbf{V}_3\rangle \\
&\quad\vdots \qquad\qquad \vdots \\
&|\mathbf{V}_{2N}\rangle \to |\mathbf{V}_N\rangle
\end{aligned}
\qquad (2.26)
$$

for the composite states of N identical atoms and the cavity, which corresponds one-to-one to the states from sequence (2.19). With this notation, the state (2.22) refers to the states $|\mathbf{V}_1\rangle, \ldots, |\mathbf{V}_N\rangle$, while the cavity-mediated photon exchange is performed between the state $|\mathbf{V}_{N+1}\rangle$ and (one of) the states $|\mathbf{V}_{N+2}\rangle, \ldots, |\mathbf{V}_{2N}\rangle$, respectively.

2.3.1 Effective single-mode Hamiltonian

Before we turn to the detailed analysis of sequence (2.19), we notice that while discussing the cavity-laser mediated interaction we considered a single-mode cavity with the frequency ω that is detuned from the atomic frequencies such that the condition (2.18) is satisfied for each atom. In practice, the cavity supports two orthogonally polarized modes which interact with each circularly polarized atom via Hamiltonian (1.39), where $\delta = \omega - \widetilde{\omega}$ is the birefringent splitting and the detuning $\Delta = (\omega_E - \omega_0) - \omega$ has been adapted for the three-level atomic configuration from Fig. 2.1(a).

It is relevant to point out that during the off-resonant interaction regime, i.e., regime in which the condition (2.18) holds true, the atomic chain interacts with both cavity modes in the same way as it would interact with a single-mode cavity characterized by the effective frequency

$$\omega_{\text{eff}} = (\omega_E - \omega_0) - \Delta_{\text{eff}}, \quad \text{where} \quad \Delta_{\text{eff}} = \left(\frac{1}{\Delta} + \frac{1}{\Delta + \delta}\right)^{-1}. \qquad (2.27)$$

In order to show this, we extend the Hamiltonian (1.39) to describe N identical atoms coupled to a cavity that supports two orthogonally polarized modes

$$\hat{H}_I^N = -\hbar \sum_{k=1}^{N} \left[i\, \frac{g(x_k)}{2} \left(\hat{\sigma}_k^\dagger \hat{a}\, e^{i\Delta t} - \hat{\sigma}_k \hat{a}^\dagger e^{-i\Delta t} \right) - \frac{g(x_k)}{2} \left(\hat{\sigma}_k^\dagger \hat{b}\, e^{i(\Delta+\delta)t} + \hat{\sigma}_k \hat{b}^\dagger e^{-i(\Delta+\delta)t} \right) \right], \qquad (2.28)$$

where $\hat{\sigma}_k = |0\rangle_k\langle e|$ and $\hat{\sigma}_k^\dagger = |e\rangle_k\langle 0|$ has been adapted for the three-level atomic configuration and $g(x_i)$ is the coupling between the cavity and i-th atom.

Now we expand in series the evolution operator associated with above Hamiltonian and keep the terms up to the second order. Then, by performing the integration and retaining only linear in time contributions, we express this evolution operator in the

2.3. Combined atom-cavity-laser interaction

form (1.41) with the effective Hamiltonian[2]

$$\hat{H}_{\text{eff}}^N = -i\,\hbar \left(\sum_{k=1}^{N} \frac{g(x_k)^2}{4\,\Delta_{\text{eff}}} |e\rangle_k \langle e| + \sum_{\substack{i,j=1 \\ (i \neq j)}}^{N} \frac{g(x_i)\,g(x_j)}{4\,\Delta_{\text{eff}}} \hat{\sigma}_i^{\dagger} \hat{\sigma}_j \right), \qquad (2.29)$$

where Δ_{eff} is the effective detuning (2.27) such that $\Delta_{\text{eff}} \geq \Delta/2$. This effective Hamiltonian governs the evolution (2.17) and it describes the (off-resonant) interaction of N two-level atoms mediated by both cavity modes. As shown in Ref. [21], however, the Hamiltonian (2.29) with replacement $\Delta_{\text{eff}} \to (\omega_a - \omega)$ governs the off-resonant interaction of N two-level atoms (ω_a) with a single-mode cavity (ω). Instead of Hamiltonian (2.28), therefore, we can describe the cavity-mediated interaction between N atoms by using the (single-mode) Hamiltonian

$$\hat{H}_I^c = -i\,\hbar \sum_{k=1}^{N} \frac{g(x_k)}{2} \left(\hat{\sigma}_k^{\dagger} \hat{c}\, e^{i\Delta_{\text{eff}} t} - \hat{\sigma}_k \hat{c}^{\dagger} e^{-i\Delta_{\text{eff}} t} \right), \qquad (2.30)$$

in which the effective cavity field is characterized by the frequency (2.27) and is associated with the rising and lowering operators $\hat{c}^{\dagger}$ and $\hat{c}$, respectively. It can be readily checked that the evolution operator (1.41) for this Hamiltonian produces the effective Hamiltonian (2.29).

Finally, we like to remind that the Hamiltonian (2.30) is expressed in the interaction picture in which the atomic and cavity Hamiltonians (1.32) and (1.13) do not enter explicitly. In many situations, however, it is convenient to consider the picture in which Hamiltonian depends explicitly on the atomic frequencies ω_E, ω_0 and the effective cavity frequency ω_{eff}. In order to switch to such a picture, we express (2.30) in the form

$$\hat{H}_I^c = \hat{H}_\bullet^c + \hat{H}_\circ^c, \quad \text{with}$$

$$\hat{H}_\bullet^c = \hbar\,\omega_{\text{eff}}\,\hat{c}^{\dagger}\hat{c} + \hbar\,\frac{N}{2}(\omega_E + \omega_0)\,\hat{I} + \hbar \sum_{k=1}^{N} \left[(\omega_E - \omega_0)\,\hat{\sigma}_k^z - i\,\frac{g(x_k)}{2}\left(\hat{\sigma}_k^{\dagger}\hat{c}\,e^{i\Delta_{\text{eff}} t} - \hat{\sigma}_k\hat{c}^{\dagger} e^{-i\Delta_{\text{eff}} t} \right) \right],$$

$$\hat{H}_\circ^c = -\hbar\,\omega_{\text{eff}}\,\hat{c}^{\dagger}\hat{c} - \hbar\,\frac{N}{2}(\omega_E + \omega_0)\,\hat{I} - \hbar\sum_{k=1}^{N} (\omega_E - \omega_0)\,\hat{\sigma}_k^z, \quad \text{where} \quad \hat{\sigma}_k^z = \frac{1}{2}(|e\rangle_k\langle e| - |0\rangle_k\langle 0|).$$

Then, we switch to the interaction picture with respect to $\hat{H}_\circ^c$, or equivalently, the Hamiltonian $\hat{H}_\bullet^c$ is transformed by means of the unitary operator $\hat{U}_\circ = \exp\left(-\frac{i}{\hbar}\hat{H}_\circ^c t\right)$.

$$\begin{aligned}
\hat{H}^c &= \hat{U}_\circ^{\dagger}\left(\hat{H}_I^c - \hat{H}_\circ^c\right)\hat{U}_\circ \\
&= \hbar\,\omega_{\text{eff}}\,\hat{c}^{\dagger}\hat{c} + \hbar\,\frac{N}{2}(\omega_E + \omega_0)\,\mathrm{I} + \hbar\sum_{k=1}^{N} \left[(\omega_E - \omega_0)\,\hat{\sigma}_k^z - i\,\frac{g(x_k)}{2}\left(\hat{\sigma}_k^{\dagger}\hat{c} - \hat{\sigma}_k\hat{c}^{\dagger} \right) \right] \\
&= \hbar\,\omega_{\text{eff}}\,\hat{c}^{\dagger}\hat{c} + \hbar\sum_{k=1}^{N} \left[(\omega_E|e\rangle_k\langle e| + \omega_0|0\rangle_k\langle 0|) - i\,\frac{g(x_k)}{2}\left(\hat{\sigma}_k^{\dagger}\hat{c} - \hat{\sigma}_k\hat{c}^{\dagger} \right) \right]. \qquad (2.31)
\end{aligned}$$

---

[2] In order to derive this Hamiltonian, we have used the fact that both cavity modes are initially empty [see sequence (2.17)].

CHAPTER 2: Interaction of three-level Λ-type atoms with cavity and laser fields

In contrast to the Hamiltonian (2.30) which depends on the effective detuning Δ_{eff}, the obtained Hamiltonian depends on the atomic frequencies ω_E, ω_0 and the effective cavity frequency ω_{eff}.

2.3.2 Combined Hamiltonian and the Schrödinger equation

For N identical atoms which move uniformly (along the x-axis) through the cavity and a laser beam and which are equally separated from each other, the evolution of coupled atoms-cavity-laser system is governed by the Hamiltonian

$$\hat{H}^{\text{tot}} = \hbar\omega_{\text{eff}}\,\hat{c}^\dagger\hat{c} + \hbar\sum_{k=1}^{N}\Big[\omega_1|1\rangle_k\langle 1| + \omega_E|e\rangle_k\langle e| + \omega_0|0\rangle_k\langle 0|$$
$$-i\left(\frac{\Omega_k(t)}{2}e^{-i\omega_L t}|e\rangle_k\langle 1| + \frac{g_k(t)}{2}\hat{c}\,|e\rangle_k\langle 0| - H.c.\right)\Big], \qquad (2.32)$$

which was obtained by combining the Hamiltonian (2.13) that governs the atom-laser interaction and the Hamiltonian (2.31) that governs the atom-cavity interaction. Notice that we considered $\varphi = 0$ in the Hamiltonian (2.13) which implies that the source of laser field should be placed at the relative position $\{0, \vartheta/k, 0\}$ from the atomic chain, where ϑ is the phase of scalar product $\boldsymbol{\epsilon}_a^* \cdot \boldsymbol{\epsilon}_J$. In the Hamiltonian (2.32), moreover, the first term describes the effective cavity energy and (the summation of) the second term describes the chain of atoms together with their time-dependent interactions with the cavity and laser fields [see (2.24)-(2.25)].

In order to simplify the analysis of evolution governed by the Hamiltonian (2.32), we switch to the interaction picture by using the unitary transformation [32, 62, 63]

$$\hat{U}_I = \exp\left[-i\,t\,\sum_{i=1}^{N}(\omega_1|1\rangle_k\langle 1| + (\omega_1+\omega_L)|e\rangle_k\langle e| + \omega_0|0\rangle_k\langle 0|) - i\,t\,\hat{c}^\dagger\hat{c}\,(\omega_L - \omega_0 + \omega_1)\right].$$

In this picture, the Hamiltonian (2.32) takes the simplified form

$$\hat{H}_I^{\text{tot}} = \hbar\,\Delta_C\,\hat{c}^\dagger\hat{c} + \hbar\sum_{k=1}^{N}\left[\Delta_L|e\rangle_k\langle e| - i\left(\frac{\Omega_k(t)}{2}|e\rangle_k\langle 1| + \frac{g_k(t)}{2}\hat{c}\,|e\rangle_k\langle 0| - H.c.\right)\right], \qquad (2.33)$$

where $\Delta_L = (\omega_E - \omega_1) - \omega_L$ and $\Delta_C = \Delta_L - \Delta_{\text{eff}}$ refer to the frequency shifts (detunings) as depicted in Fig. 2.1(b). The time evolution governed by this Hamiltonian is described by the Schrödinger equation

$$i\hbar\frac{d|\Psi(t)\rangle}{dt} = \hat{H}_I^{\text{tot}}|\Psi(t)\rangle \quad \text{with the ansatz}\quad |\Psi(t)\rangle = \sum_{i=0}^{2N}C_i(t)|\mathbf{V}_i\rangle, \qquad (2.34)$$

where $C_0(t),\ldots,C_{2N}(t)$ are the complex and normalized amplitudes such that $C_k(0) = \delta_{k1}$. For this ansatz, the Schrödinger equation (2.34) gives rise to a closed system of

$2N+1$ equations

$$i\dot{C}_0(t) = \Delta_C\, C_0(t) + \frac{i}{2}\sum_{k=1}^{N} g_k(t)\, C_{N+k}(t), \qquad (2.35\text{a})$$

$$i\dot{C}_k(t) = \frac{i}{2}\Omega_k(t)\, C_{N+k}(t), \qquad (2.35\text{b})$$

$$i\dot{C}_{N+k}(t) = \Delta_L\, C_{N+k}(t) - \frac{i}{2}\Big(\Omega_k(t)\, C_k(t) + g_k(t) C_0(t)\Big), \qquad (2.35\text{c})$$

where $k = 1, \ldots, N$ and the dot denotes the time derivative. This system of equations describes the evolution of coupled atoms-cavity-laser system that is governed by the Hamiltonian (2.32) and that corresponds to the situation in which N identical (three-level) atoms are moving uniformly through the cavity and a laser beam.

2.4 Far off-resonant interaction regime

As we explained in the previous section, the $N+1$ states $|\mathbf{V}_0\rangle$ and $|\mathbf{V}_{N+1}\rangle, \ldots, |\mathbf{V}_{2N}\rangle$ remain (almost) unpopulated if the atom-cavity and atom-laser detuning satisfy the conditions (2.18) and (2.20), respectively. However, since the Hamiltonian (2.33) contains two detunings Δ_L and Δ_C at the same time, these two conditions must be supplemented with one more condition

$$\Delta_L \Delta_C \gg g_i(t)\, g_j(t), \quad i,j = 1, \ldots, N. \qquad (2.36)$$

In order to separate the evolution of states $|\mathbf{V}_0\rangle$ and $|\mathbf{V}_{N+1}\rangle, \ldots, |\mathbf{V}_{2N}\rangle$ from Eqs. (2.35), we apply the adiabatic elimination procedure [95] which assumes an adiabatic behavior of amplitudes $C_0(t)$ and $C_{N+1}(t), \ldots, C_{2N}(t)$, and to a good approximation therefore, vanishing of the time derivatives associated to these amplitudes.

First, we exploit the derivative $\dot{C}_0(t) \cong 0$ and obtain with help of Eq. (2.35a) the equation

$$C_0(t) = -\frac{i}{2\Delta_C}\sum_{k=1}^{N} g_k(t)\, C_{N+k}(t). \qquad (2.37)$$

By inserting this equation in Eq. (2.35c) together with the time-derivatives $\dot{C}_{N+k}(t) \cong 0$ ($k = 1, \ldots, N$), we obtain the set of equalities

$$\sum_{k=1}^{N}\left(\delta_{ki} - \frac{g_k(t)\, g_i(t)}{4\Delta_C \Delta_L}\right) C_{N+k}(t) - i\,\frac{\Omega_j(t)}{2\Delta_L}\, C_j(t) \qquad (2.38)$$

which we readily express in the matrix form $\mathbf{A}\vec{\mathbf{X}} = \vec{\mathbf{C}}$ with $\vec{\mathbf{X}} = (C_{N+1}(t), \ldots, C_{2N}(t))^T$, $\vec{\mathbf{C}} = \frac{i}{2\Delta_L}\,(\Omega_1(t)\, C_1(t), \ldots, \Omega_N(t)\, C_N(t))^T$, and where the matrix elements of $\mathbf{A}$ are given by[3]

$$A_{ij} = -\frac{g_i(t)\, g_j(t)}{4\Delta_C \Delta_L},\ (i \neq j)\quad \text{and}\quad A_{ii} = 1 - \frac{g_i(t)^2}{4\Delta_C \Delta_L} \cong 1. \qquad (2.39)$$

---
[3] Here, for the diagonal matrix elements A_{ii} we used the condition (2.36).

CHAPTER 2: Interaction of three-level Λ-type atoms with cavity and laser fields

The amplitudes $C_{N+1}(t), \ldots, C_{2N}(t)$ which corresponding to the vector $\vec{X}$ are found by inverting the matrix $\mathbf{A}$

$$\vec{X} = \mathbf{A}^{-1}\vec{C} = \frac{\text{adj}[\mathbf{A}]}{\det[\mathbf{A}]}\vec{C}, \qquad (2.40)$$

where $\text{adj}[\mathbf{A}]$ and $\det[\mathbf{A}]$ denote the adjugate matrix and the determinant, respectively. It can be readily shown, moreover, that $\det[\mathbf{A}]$ and the matrix elements of $\text{adj}[\mathbf{A}]$ take the form

$$\det[\mathbf{A}] = 1 - \sum_{\substack{i,j=1 \\ (i\neq j)}}^{N} \frac{g_i(t)^2 \, g_j(t)^2}{(4\Delta_L \Delta_C)^2} - \sum_{\substack{i,j,k=1 \\ (i\neq j\neq k)}}^{N} \frac{g_i(t)^2 \, g_j(t)^2 \, g_k(t)^2}{(4\Delta_L \Delta_C)^3} - \cdots$$

$$(\text{adj}[\mathbf{A}])_{ij} = \frac{g_i(t) \, g_j(t)}{(4\Delta_L \Delta_C)^{N-1}} \prod_{\substack{k=1 \\ (k\neq i\neq j)}}^{N} \left(4\Delta_L \Delta_C + g_k(t)^2\right), \quad i \neq j$$

$$(\text{adj}[\mathbf{A}])_{kk} = 1 - \sum_{\substack{i,j=1 \\ (i\neq j\neq k)}}^{N} \frac{g_i(t)^2 \, g_j(t)^2}{(4\Delta_L \Delta_C)^2} - \sum_{\substack{i,j,m=1 \\ (i\neq j\neq m\neq k)}}^{N} \frac{g_i(t)^2 \, g_j(t)^2 \, g_m(t)^2}{(4\Delta_L \Delta_C)^3} - \cdots$$

which, due to the condition (2.36), reduce to the expressions

$$\det[\mathbf{A}] \cong 1; \quad (\text{adj}[\mathbf{A}])_{ij} \cong \frac{g_i(t) \, g_j(t)}{4\Delta_C \Delta_L} \quad (i\neq j); \quad (\text{adj}[\mathbf{A}])_{kk} \cong 1. \qquad (2.41)$$

Owning to these simplified expressions, the matrix equation (2.40) implies ($k = 1, \ldots, N$)

$$C_{N+k}(t) = i\frac{\Omega_k(t)}{2\Delta_L}C_k(t) + i\sum_{\substack{j=1 \\ (j\neq k)}}^{N} \frac{g_k(t) \, g_j(t) \, \Omega_j(t)}{8\Delta_C \Delta_L^2}C_j(t). \qquad (2.42)$$

By inserting this equation in Eq. (2.35b), we obtain the set of differential equations

$$i\dot{C}_k(t) = -\frac{\Omega_k(t)^2}{4\Delta_L}C_k(t) - \sum_{\substack{j=1 \\ (j\neq k)}}^{N} \frac{g_k(t) \, \Omega_k(t) \, g_j(t) \, \Omega_j(t)}{16\Delta_C \Delta_L^2}C_j(t), \quad (k=1,\ldots,N) \qquad (2.43)$$

which contain only amplitudes $C_1(t), \ldots, C_N(t)$ corresponding to vectors $|\mathbf{V}_1\rangle, \ldots, |\mathbf{V}_N\rangle$.

2.4.1 Effective Hamiltonian and the asymptotic coupling

The Eqs. (2.43) which we derived from Eqs. (2.35) by using the adiabatic elimination procedure, in turn, can be derived directly from the Schrödinger equation

$$i\hbar\frac{d|\Phi(t)\rangle}{dt} = \hat{H}^{\text{tot}}_{\text{eff}}|\Phi(t)\rangle \quad \text{with the ansatz} \quad |\Phi(t)\rangle = \sum_{i=1}^{N}C_i(t)|\mathbf{V}_i\rangle, \qquad (2.44)$$

40

2.4. Far off-resonant interaction regime

which is associated with the effective Hamiltonian

$$\hat{H}_{\text{eff}}^{\text{tot}} = -\hbar \sum_{k=1}^{N} \frac{\Omega_k(t)^2}{4\,\Delta_L} |\mathbf{V}_k\rangle\langle\mathbf{V}_k| - \hbar \sum_{\substack{i,j=1 \\ (i\neq j)}}^{N} \frac{q_i(t)\,q_j(t)}{2\,\Delta_C\,\Delta_L^2} \left(\hat{S}_i^\dagger \hat{S}_j + \hat{S}_i \hat{S}_j^\dagger \right). \quad (2.45)$$

In this Hamiltonian, moreover, $\hat{S}_i^\dagger = |1\rangle_i\langle 0|$ and $\hat{S}_i = |0\rangle_i\langle 1|$ denote the atomic two-photon excitation and de-excitation operators, respectively, and where we introduced the combined coupling frequency

$$q_i(t) = g_i(t)\,\Omega_i(t)/4 = q_\circ \exp\left[-(x_i^o + v\,t)^2/w_C^2\right], \quad i = 1,\ldots,N \quad (2.46)$$

with $w_C = w_L\,w/\sqrt{w^2 + w_L^2}$ being the waist of combined atom-laser fields and $q_\circ = \Omega_\circ\,g_\circ/4$ being the combined vacuum Rabi splitting. In contrast to the Hamiltonian (2.33), the Hamiltonian (2.45) describes the sequence (2.21) that governs the atomic evolution which is mediated by the combined laser and cavity fields. In order to summarize the steps before, we found that the evolution of N three-level atoms is reduced to the evolution of effectively two-level atoms which interact with each other via a two-photon exchange such that the atomic excited states are adiabatically eliminated and remains (almost) unpopulated [63].

Recall that in our setup, the atomic chain is initialized outside the cavity and transported through the cavity with a constant velocity as displayed in Fig. 2.2. The entangled state (2.22) is generated after the atomic chain leaves the cavity and decouples from both the cavity and laser fields. In order to find the atomic evolution governed by the Hamiltonian (2.45), therefore, we need to integrate the Schrödinger equation (2.44) inside the time period in which the cavity-laser mediated interaction between the atoms is switched on and off, respectively. In a high finesse cavity, however, the Gaussian profile (2.24) approximates quite well the cavity field and, therefore, we integrate (2.44) from $-\infty$ to $+\infty$

$$|\Phi(\infty)\rangle = \exp\left[-\frac{i}{\hbar}\int_{-\infty}^{\infty} \hat{H}_{\text{eff}}^{\text{tot}}\,dt\right] |\mathbf{V}_1\rangle = \sum_{i=1}^{N} C_i(\infty)|\mathbf{V}_i\rangle \quad (2.47)$$

in the same way as we calculated effective atom-cavity interaction time in section 1.2.2.

It can be readily checked that the integral $-\frac{1}{\hbar}\int_{-\infty}^{\infty} \hat{H}_{\text{eff}}^{\text{tot}}\,dt$ is given by the expression

$$\sqrt{\frac{\pi}{2}}\frac{\Omega_\circ^2\,w_L}{4\,\Delta_L\,v}\sum_{k=1}^{N}|\mathbf{V}_k\rangle\langle\mathbf{V}_k| + \sqrt{\frac{\pi}{2}}\frac{q_\circ^2\,w_C}{\Delta_C\,\Delta_L^2\,v}\sum_{\substack{i,j=1 \\ (i\neq j)}}^{N} \exp\left[-\frac{(|i-j|\,d)^2}{2\,w_C^2}\right]|1_i,0_j\rangle\langle 0_i,1_j|, \quad (2.48)$$

which depends on the two parameters v and d since the frequency shifts: Δ_L, Δ_C, coupling constants: $g_\circ$, $\Omega_\circ$, and waists: w_L, w_C are fixed by a particular experimental setup. The time-independent atomic wave-function $|\Phi(\infty)\rangle$ along with amplitudes

CHAPTER 2: Interaction of three-level Λ-type atoms with cavity and laser fields

$C_i(\infty)$, therefore, depend only on the (v,d) pair and the wave-function (2.47) can be expressed in the form

$$|\Phi(v,d)\rangle = \exp\left[i\,\hat{M}\right]|\mathbf{V}_1\rangle = \sum_{i=1}^{N} C_i(v,d)|\mathbf{V}_i\rangle, \qquad (2.49)$$

where the matrix elements $M_{ij} = \langle \mathbf{V}_i|\hat{M}|\mathbf{V}_j\rangle$ are given by

$$M_{ii} = \sqrt{\frac{\pi}{2}}\,\frac{\Omega_\circ^2\,w_L}{4\,\Delta_L\,v} \quad \text{and} \quad M_{ij} = \theta\left(v,|i-j|\,d\right) \quad \text{for} \quad i \neq j, \qquad (2.50)$$

with

$$\theta(v,d) = \sqrt{\frac{\pi}{2}}\,\frac{q_\circ^2\,w_C}{\Delta_C\,\Delta_L^2\,v}\,\exp\left[-\frac{d^2}{2\,w_C^2}\right]. \qquad (2.51)$$

The expression (2.51) can be interpreted as the asymptotic coupling for a pair of atoms which move with the same velocity v and are separated by the distance d from each other. Moreover, the amplitudes of wave-function (2.49)

$$C_i(v,d) = \langle \mathbf{V}_i|\exp\left[i\,\hat{M}\right]|\mathbf{V}_1\rangle = (\exp\left[i\,\mathbf{M}\right])_{i1} \qquad (2.52)$$

can be computed routinely, for instance, by diagonalization of the matrix $\mathbf{M}$ that is characterized by the elements (2.50). In chapter 6, we shall discuss the properties of $|\Phi(v,d)\rangle$ for different values of N and shall display those v and d pairs, for which this wave-function reduces to the W state (2.23).

2.5 Summary

In this chapter, we first described the interaction of a single atom with monochromatic light field in the semiclassical regime. We found that the evolution of atomic state that interacts (resonantly) with a laser field is given by the expressions

$$|e\rangle \to \cos(\Omega_\circ t/2)\,|e\rangle + \sin(\Omega_\circ t/2)\,e^{i\varphi}\,|g\rangle, \qquad (2.53a)$$

$$|g\rangle \to \cos(\Omega_\circ t/2)\,|g\rangle - \sin(\Omega_\circ t/2)\,e^{-i\varphi}\,|e\rangle, \qquad (2.53b)$$

where angle φ is set by the relative position of atom with respect to the field source.

Furthermore, we explained our scheme for generation of atomic multipartite entangled states which is based on the off-resonant interaction regime of three-level atoms placed inside the cavity and coupled simultaneously to a laser beam [62, 63]. We found that the cavity and a laser beam mediate together the interaction between atoms which are simultaneously coupled to them. By performing the adiabatic elimination procedure, we showed that the evolution of initially uncorrelated atoms is described by the sequence (2.21) and is governed by the Hamiltonian (2.45). According to this Hamiltonian, moreover, the evolution of three-level atoms is reduced to the evolution of effectively two-level atoms which interact with each other via a two-photon exchange such that the fast decaying atomic excited states remain almost unpopulated. This energy exchange is quantitatively described by the W-class state (2.49) and is characterized by the complex amplitudes (2.52) which, in turn, are determined by the atomic velocities and inter-atomic distances. The asymptotic coupling (2.51) tells explicitly how these amplitudes depends on these two parameters. By setting appropriate velocities of atoms and inter-atomic distances, therefore, one can generate the entangled W state (2.23) from the W-class state (2.22) after the atomic chain leaves the cavity and decouples from both cavity and laser fields.

We finally mention that the off-resonant interaction regime, is robust with regard to decoherence effects since the cavity mode and (fast decaying) excited atomic states remain almost unpopulated during the entire evolution. This robustness, in turn, plays one crucial role in the generation of multipartite entangled W states between atomic qubits encoded into the level structure of neutral atoms, and where the cavity plays the role of a data bus that mediates the interaction between these atomic qubits.

Part II

Cavity-QED experimental setups

Chapter 3

Microwave cavity setup

In the first chapter, we investigated the situation in which a single two-level atom is coupled to a cavity field via dipole-field interaction such that atomic transition frequency matches one of resonant modes of cavity. We found that the evolution of coupled atom-cavity system yields a coherent exchange of energy between the constituents which describes a time-varying entanglement of the atomic and cavity photon field states. Owning to this time-varying entanglement, we concluded that by setting an appropriate interaction time and an initial (uncorrelated) atom-cavity state, one can generate multipartite entangled states of atomic qubits which pass sequentially through the cavity.

In order to apply these effects in practice, however, the vacuum Rabi splitting $g_\circ$ – the position-independent part of atom-field coupling (1.38), should be much larger than (i) atomic spontaneous emission rate γ, (ii) cavity relaxation rate κ, and (iii) reciprocal of the atom-cavity interaction time as required for realization of any particular atom-cavity evolution(s). The conditions (i) and (ii), roughly speaking, define the (so-called) *strong coupling* regime of the atom-cavity interaction which ensures that the energy exchange between the constituents is reversible and it develops faster than photon loss due to the cavity relaxation or atomic decay. The reversibility of energy exchange, in turn, ensures that the atom-cavity system undergoes the Rabi oscillations (time-varying entanglement of atomic and cavity photon states) and which is widely exploited in the third part of this manuscript.

The definition (1.38) implies that the vacuum Rabi splitting is determined by the atomic dipole momentum, cavity mode frequency, and the cavity mode volume. In order to achieve the strong coupling regime, therefore, one should consider an atom which possesses a larger atomic dipole moment or decrease the cavity mode volume. Since the atomic dipole is essentially determined by the separation of valence electron from its nucleus, one can choose the Rydberg atoms which are highly excited (alkali) atoms with the principal quantum number n of the order of 60. Since the atomic dipole of a Rydberg atom grows with n^2 and the transition between its two neighboring levels

CHAPTER 3: Microwave cavity setup

lies in the microwave domain, such an atom exhibits a strong coupling to the microwave light field. By placing such Rydberg atom inside a microwave resonator, therefore, the strong coupling regime can be experimentally achieved. The groups of H. Walther in Garching [33] and S. Haroche in Paris [34] have capitalized on this combination and developed experimental setups based on Rydberg atoms and microwave cavities. In this chapter, we shall describe in details the basic constituents of setup developed in the group of S. Haroche.

3.1 Microwave cavity

A microwave cavity is the heart of setup that we exploit in this manuscript. This cavity is an open resonator that consists of two polished spherical niobium mirrors facing each other and where each mirror has a diameter of 50 mm and a radius of curvature $R_M = 40$ mm. The cavity supports resonant frequency $\omega \approx 2\pi \cdot 51.1$ GHz which is about the frequency of transition between the (Rydberg) states 50 and 51 of a rubidium atom (see below). The mirrors are separated by the distance $L = 27$ mm (in the origin of $x-y$ plane) and accommodate $k = 9$ antinodes along the cavity axis such that $L \approx k\lambda/2$ and where $\lambda = 2\pi c/\omega \approx 5.9$ mm is the wavelength associated with the resonant cavity mode. By developing new coating and polishing techniques of cavity mirrors, moreover, the cavity relaxation time $1/\kappa \simeq 130$ ms has been achieved in the group of S. Haroche [71].

The transverse cavity field components are described by the cavity field structure (1.27) that involves the width of transverse Gaussian profile $W(z)$ and the radius of curvature of the light wavefront $R(z)$ [see (1.26)]. The cavity mode waist w can be calculated from the condition $R_M = R(z_+) = R(z_-)$ which leads to the expression

$$w = \left(\frac{\lambda L}{2\pi} \sqrt{\frac{2 R_M}{L} - 1} \right)^{1/2}. \tag{3.1}$$

The cavity waist, therefore, is completely determined by the curvature R_M, distance L, and resonant wavelength λ. In the group of S. Haroche, these parameters have been chosen such that the cavity waist is nearly equal to the cavity resonant wavelength, i.e., 5.9 mm.

The cavity polarization vector ϵ lies in the $x-y$ plane [see Fig. 1.2(b)] and it supports two modes (C_x and C_y) with orthogonal linear polarizations. With perfectly spherical mirrors, moreover, the symmetry of cavity shape ensures that these two modes are strictly degenerate. In practice, however, various imperfections of cavity mirrors produce a birefringent splitting that lifts this degeneracy by the value of $\delta = 2\pi \cdot 128$ KHz. In addition, the mode volume is obtained by integrating (1.27) over the space, which yields

$V = \pi w^2 L/4 \approx 740\,\text{mm}^3$. According to the expression (1.15), in turn, the amplitude of vacuum field in the cavity center $[f(\mathbf{r}') = 1]$ is given by the expression

$$\mathcal{E} = \sqrt{\frac{\hbar\omega}{2\,\varepsilon_0\,V}} \approx 1.5 \cdot 10^{-3}\,\text{Volt/m}. \tag{3.2}$$

This result is of the order of a millivolt per meter and implies that the vacuum field in the cavity center has the magnitude of a macroscopic field.

Since the resonant cavity frequency lies in the microwave domain, the cavity mode contains residual thermal photons. In order to realize the atom-cavity energy exchange as described in previous part of this manuscript, the mean number of thermal photons should be negligible. However, it can be checked by substituting the cavity frequency ω in the expression [82]

$$n_\text{th} = \frac{1}{e^{\beta\hbar\omega} - 1}, \quad \text{where} \quad \beta = (k_B\,T)^{-1}, \tag{3.3}$$

that the cavity mirrors should be cooled down to $T \approx 1\,\text{K}$ in order to satisfy the condition $n_\text{th} \ll 1$ as necessary for the cavity-QED experiments. In the group of S. Haroche, the cavity was cooled down to 0.8 K which yields an average number of $n_\text{th} \approx 0.05$ thermal photons and ensures that the contribution of residual thermal photons can be neglected.

3.2 Circular Rydberg atoms as qubits

In contrast to the *stationary* photonic qubits which are associated to the microwave cavity, each Rydberg atom encodes one *flying* qubit which can be strongly coupled to the cavity field due to its large dipole momentum. A circular Rydberg state [83, 84] is a highly excited alkali atom (atoms of rubidium in the experiments by S. Haroche), in which the single valence electron is excited such that $\ell = |m| = n - 1$, where n, ℓ, and m are the principal, orbital, and magnetic quantum numbers, respectively. A circular Rydberg state, therefore, is determined by its quantum number n, and shall be referred below as $|n_c\rangle$.

The energy of a circular state is well described by the hydrogen-like expression $E_n = -R_\text{rb}/n^2$, where R_rb is the Rydberg constant for rubidium. This energy, in turn, determines the frequency of atomic $|n_c\rangle \to |(n-1)_c\rangle$ transition between two neighboring levels

$$\omega_n = \frac{1}{\hbar}(E_{n-1} - E_n) = \frac{R_\text{rb}}{\hbar}\frac{1}{n^2}\left[\left(1 - \frac{1}{n}\right)^{-2} - 1\right] \approx 2\frac{R_\text{rb}}{\hbar}\frac{1}{n^3}, \tag{3.4}$$

which corresponds to the frequency in the few tens of GHz range for values $n = 20,\ldots,60$. The same atomic transition, moreover, is characterized by the (circular) transition polarization $\boldsymbol{\epsilon}_a^+$ and by the dipole matrix element [35]

$$d_n = a_0\,q\,n^2/\sqrt{2}. \tag{3.5}$$

CHAPTER 3: Microwave cavity setup

Obviously, the energy E_n mentioned above does not account for contributions due to the (hyper)fine atomic structure. The fine structure, however, scales as $1/n^5$ and it is only a few hundred hertz for values $n = 20, \ldots, 60$. These contributions, therefore, are very small in circular states and can be safely neglected.

In classical terms, furthermore, the orbit of a valence electron around the core is a circle with radius $a_0 n^2$, where a_0 is the Bohr radius. Among all possible bound orbits, the circular ones have the smallest average acceleration and the electron always remains far from the core. These features lead to one minimum loss of radiation and, therefore, to a longer radiative lifetime which is crucial for the cavity-QED experiments. The spontaneous emission rate γ_n of a Rydberg atom is given by the expression [35]

$$\gamma_n = \frac{4}{3} \frac{R_{\rm rb}\,\alpha^3}{\hbar\, n^5}, \qquad (3.6)$$

where α is the fine structure constant.

The described properties of Rydberg atoms imply that high values of n are preferable for the experimental perspective since the atomic dipole d_n and lifetime $1/\gamma_n$ scale as n^2 and n^5, respectively. According to the expression (3.4), however, high values of n imply smaller frequencies which, in turn, lead to more thermal photons if the cavity resonant frequency coincides with ω_n [see (3.3)]. A compromise between these requirements has been found in the group of S. Haroche, and namely, the Rydberg states with $n = 49$, 50, and 51 have been chosen. With this choice, the atom exhibits a huge dipole matrix element $1767\,q\,a_0$ and the atomic $|50_c\rangle \leftrightarrow |51_c\rangle$ transition at 51.099 GHz matches the resonant frequency of cavity mode C_x. Moreover, the circular Rydberg atoms can travel a few meters within their lifetimes $1/\gamma \simeq 36$ ms and, therefore, the spontaneous emission can be neglected in an experimental setup with the size of about few tens of centimeters and typical atomic velocities of about 500 m/s. Finally, the transition $|49_c\rangle \leftrightarrow |50_c\rangle$ at 54.3 GHz is far off-resonant with the cavity mode and is unaffected by the atom-cavity coupling and corresponding time-evolution.

3.2.1 Ramsey plates

In the previous section, we mentioned that the microwave cavity is compatible with the atomic $|50_c\rangle \leftrightarrow |51_c\rangle$ transition. By this choice, therefore, the Rydberg states $|50_c\rangle$ and $|51_c\rangle$ encode one (flying) qubit that interacts with one (stationary) qubit encoded in the cavity states $\{|0\rangle, |1\rangle\}$ or $\{|\bar{0}\rangle, |\bar{1}\rangle\}$. For brevity, the states $|51_c\rangle$, $|50_c\rangle$, and $|49_c\rangle$ shall be referred below as excited state $|e\rangle$, ground state $|g\rangle$, and auxiliary state $|a\rangle$, respectively. Each Rydberg atom emitted from the atomic source, moreover, is prepared in one of states $|e\rangle$ or $|g\rangle$ by using the procedure from Ref. [83] and has a constant velocity along the axis of experimental setup [see Fig. 1.4]. It is essential, however, to

3.2. Circular Rydberg atoms as qubits

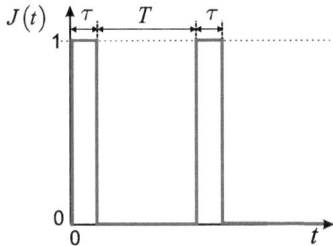

Figure 3.1: Temporal sequence of two short (off-resonant) Ramsey pulses to realize the rotation of atomic state with a non-zero angle $\varphi = \Delta_J T$. See text for explanation.

prepare an atomic qubit in the state

$$\cos(\phi/2)\,|e\rangle + \sin(\phi/2)\,e^{i\varphi}\,|g\rangle \qquad (3.7)$$

that is characterized by two arbitrary angles ϕ and φ. These two angles, moreover, can be interpreted as the polar angles in three-dimensional spherical coordinates and lead to the representation of above state as a point on a unit sphere – the Bloch sphere (see section 5.2).

In section 2.1, furthermore, we demonstrated that the state (3.7) can be realized by acting on the atom by a classical field (laser beam) that is tuned in resonance with the atomic transition. According to the Eq. (2.53a), therefore, the state (3.7) is realized by setting an appropriate interaction time $t = \phi/\Omega_\circ$ and the phase φ, while the atom (prepared in the excited state) passes through the laser beam. In contrast to the angle ϕ that is set by the atom-field interaction time, the angle φ is set by the relative position $\mathbf{r}' = \{0, -(\varphi+\vartheta)/k, 0\}$ of atom with respect to field source, where ϑ is the phase of scalar product $\boldsymbol{\epsilon}_a^* \cdot \boldsymbol{\epsilon}_J$ [see (2.4)]. From the practical perspective, therefore, the manipulation of φ is feasible but complicated to realize in practice since it requires adjusting of the relative position of laser during the experiment. In this section, we describe another procedure to manipulate the angle φ that is based on the short off-resonant pulses of classical field.

In contrast to section 2.1 in which we considered a laser beam characterized by the frequency ω_L and the atom-laser coupling (2.5), in this section we consider the classical field produced by a microwave field generator (field source) and injected between the Ramsey plates. The Ramsey plates consist of two low quality mirrors such that the relaxation time (of field injected between these mirrors) lies in the nanosecond range and is much smaller than the relaxation time of cavity introduced in section 3.1. This small relaxation time, therefore, ensures that this microwave field does not produce any

CHAPTER 3: Microwave cavity setup

entanglement with an atom [85] and thus the semiclassical approximation can be applied to describe the interaction of such field with an atom. The microwave field, furthermore, is characterized by the frequency ω_J and the atom-field coupling $\Omega_\circ = 2\,\mathrm{d}\,\mathcal{E}_\circ\,|\epsilon_a^* \cdot \epsilon_J|/\hbar$, where $\mathcal{E}_\circ$ and ϵ_J are the (real) field amplitude and polarization of microwave field, respectively. We assume, moreover, that the shape and size of Ramsey plates are chosen such that the atom-field Hamiltonian (2.4) becomes

$$\hat{H}_I^s = -i\,\hbar\,\frac{\Omega_\circ}{2}\left(\hat{\sigma}^\dagger e^{i\,\Delta_J t} - \hat{\sigma}\,e^{-i\,\Delta_J t}\right) J(t), \qquad (3.8)$$

where $\Delta_J = \omega_a - \omega_J$ is the atom-field detuning, $\hat{\sigma} = |g\rangle\langle e|$ and $\hat{\sigma}^\dagger = |e\rangle\langle g|$ denote the excitation lowering and rising operators, respectively, and where $J(t) = 1$ when the field source is switched on and $J(t) = 0$ otherwise.

In the resonant regime $\Delta_J = 0$, the above Hamiltonian describes the (unitary) rotation $\hat{R}(\Omega_\circ t, 0)$ of the atomic state, where $\hat{R}(\phi, \varphi)$ is given by the matrix [see (2.53)]

$$\hat{R}(\phi,\varphi) = \begin{pmatrix} \cos(\phi/2) & -\sin(\phi/2)\,e^{-i\varphi} \\ \sin(\phi/2)\,e^{i\varphi} & \cos(\phi/2) \end{pmatrix} \qquad (3.9)$$

expressed in the basis $\{|g\rangle, |e\rangle\}$. In order to implement the rotation of atomic state with a non-zero angle φ, moreover, we choose the time-dependent function $J(t)$ as displayed in Fig. 3.1 and which implies two atom-field interaction periods (Ramsey pulses) of duration τ being separated by the time delay $T \gg \tau$. During these interaction times, the atom-field detuning is slightly off-resonant and thus the atom-field evolution during each pulse is given by

$$\exp\left[-\frac{\Omega_\circ}{2}\int_0^\tau \left(\hat{\sigma}^\dagger e^{i\,\Delta_J t} - \hat{\sigma}\,e^{-i\,\Delta_J t}\right) dt\right], \quad \exp\left[-\frac{\Omega_\circ}{2}\int_0^\tau \left(\hat{\sigma}^\dagger e^{i\,\Delta_J (t-T)} - \hat{\sigma}\,e^{-i\,\Delta_J (t-T)}\right) dt\right].$$

Now we assume that atom-field detuning Δ_J is much smaller than the spectral width $1/\tau$ such that the variation $\Delta_J t$ can be neglected for each interaction pulse. With this assumption in mind, the above evolutions become

$$\exp\left[-\frac{\Omega_\circ}{2}\left(\hat{\sigma}^\dagger - \hat{\sigma}\right)\tau\right], \quad \exp\left[-\frac{\Omega_\circ}{2}\left(\hat{\sigma}^\dagger e^{-i\,\Delta_J T} - \hat{\sigma}\,e^{i\,\Delta_J T}\right)\tau\right]$$

and correspond to the Hamiltonian (2.6) with $\varphi = 0$ and $\varphi = \Delta_J T$, respectively. We can conclude, therefore, that the sequence of two short (off-resonant) pulses shown in Fig. 3.1 produces the rotation $\hat{R}(\Omega_\circ \tau, \Delta_J T) \cdot \hat{R}(\Omega_\circ \tau, 0)$ of atomic state, and where the time delay T is controlled by switching on and off the source of microwave field while the atom passes through the Ramsey plates.

We just explained that a single resonant Ramsey pulse realizes the rotation (i) $\hat{R}(\alpha, 0)$ with the angle $\alpha = \Omega_\circ \tau$ determined by the atom-field interaction time τ, while two off-resonant pulses realize the rotation (ii) $\hat{R}(\beta, \varphi) \cdot \hat{R}(\beta, 0)$ with angles $\beta = \Omega_\circ \tau$ and $\varphi = \Delta_J T$ determined by the atom-field interaction time τ, delay T, and detuning Δ_J

such that the condition $\Delta_J \tau \ll 1$ must be satisfied. Moreover, it is possible to apply sequentially the pulses (i) and (ii) upon the same atom and within the same Ramsey plates in order to realize the rotation

$$\hat{R}(\beta,\varphi) \cdot \hat{R}(\beta,0) \cdot \hat{R}(\alpha,0) = \hat{R}(\beta,\varphi) \cdot \hat{R}(\phi,0), \quad \text{where} \quad \phi = \alpha + \beta. \tag{3.10}$$

By setting appropriately the angle α, therefore, it is possible to cancel the term $\hat{R}(\phi,0)$ in (3.10) and realize only the rotation $\hat{R}(\beta,\varphi)$. We can conclude, therefore, that the sequence of three Ramsey pulses (one resonant and two off-resonant) realizes the rotation (iii) $\hat{R}(\beta,\varphi)$. Finally, we mention that the classical microwave field produced between the Ramsey plates can drive also $g \leftrightarrow a$ atomic transition in order to generate a superposition between states $|g\rangle$ and $|a\rangle$. In this case, the rotation of atomic state is describes by the matrix (3.9) expressed in the basis $\{|a\rangle, |g\rangle\}$ and the above rotations (i), (ii), and (iii) are realized by the same pulses.

3.3 Summary

A chain of circular Rydberg atoms and a microwave cavity are the main ingredients of our schemes for generation of various entangled states which we present in the third part of this manuscript. Recall that in order to exploit in practice the atom-cavity entanglement mechanism, each atom from the chain should be strongly coupled to the cavity such that the energy exchange between them is reversible and it develops faster than the photon loss due to the cavity relaxation and/or atomic decay, i.e., both the conditions $g_\circ \gg \kappa$ and $g_\circ \gg \gamma$ must be satisfied at the same time. In this chapter, we mentioned that the rates associated with atomic decay and cavity relaxation are $\kappa/2\pi = 4.46$ Hz and $\gamma/2\pi = 1.23$ Hz, respectively. At the same time, the vacuum Rabi frequency at the cavity center (where the cavity field amplitude takes its maximum) is $g_\circ/2\pi = 50$ KHz and it corresponds to a Rabi period $2\pi/g_\circ = 20$ μs. We can conclude, therefore, that the mentioned conditions for the strong coupling regime are largely satisfied and this is possible merely due to the ultra-high quality factor of cavity mirrors and the circular state transition of a Rydberg atom.

We also mention that a series of striking experiments has been performed and reported in the group of S. Haroche. For instance, a direct test of field quantization in a cavity [37], generation of a maximally entangled state with an atomic pair [38, 39], realization of a two-qubit phase gate [40], non-destructive measurement of a single photon [43, 44], generation of entangled atomic triplet [41], generation of Schrödinger cat states along with exploration of their decoherence dynamics [45, 46], and e.t.c. Obviously, any practical realization of theoretical schemes based on the resonant atom-cavity interaction suffers from various practical limitations: (i) imperfect Ramsey and Rabi pulses,

CHAPTER 3: Microwave cavity setup

(ii) cavity relaxation and residual thermal photons, (iii) low efficiency of atomic source and ionization detector, to name just a few of them. Various improvements, therefore, are further needed in order to overcome these difficulties and achieve an excellent control of the atom-cavity and (cavity mediated) atom-atom entanglement in the framework of microwave cavity-QED.

Chapter 4

Optical cavity setup

In the second chapter, we investigated the situation in which a chain consisting of three-level atoms is coupled to a cavity field and a laser beam such that atomic transition frequencies are off-resonant with respect to both (detuned) cavity modes. We found that the evolution of coupled atom-cavity-laser system implies a coherent two-photon exchange between the atoms which describes a W-class state that is generated between the qubits encoded in the ground and metastable states of each three-level atom. Owning to this W-class state, moreover, we concluded that by setting an appropriate velocity of atomic chain, inter-atomic distance, and an initial (uncorrelated) atom-cavity state, one can generate entangled W states right after the atomic chain has passed through the cavity.

As we mentioned in the third chapter, however, in order to apply these effects in practice the vacuum Rabi splitting $g_\circ$ should be much larger than atomic spontaneous emission rate γ and cavity relaxation rate κ. These two conditions define the strong coupling regime of atom-cavity interaction and ensure that the cavity mediated energy exchange between the atoms is realized faster than photon loss due to the cavity relaxation or atomic decay. The cavity mediated energy exchange, in turn, is the main mechanism that generates the W-class state. The definition (1.38) implies, moreover, that in order to achieve strong coupling regime one should consider an atom which possesses a larger atomic dipole moment or decrease the cavity mode volume. In contrast to the previous chapter in which we utilized Rydberg atoms with large dipole momentum, in this chapter we describe the second approach in which the small dipole momentum is compensated by a reasonably small cavity volume.

Since a cavity with smaller volume produces higher amplitude of intracevity vacuum field, an atom with low-lying electronic states can still exhibit a strong coupling to the light field. Low-lying electronic states, moreover, imply that the atomic transition between two neighboring levels lies in the optical domain. By placing such an atom inside the resonator that supports optical resonant frequency, therefore, the strong coupling regime can be experimentally achieved and the scheme developed in the second chapter can be applied. The groups of J. Kimble in Pasadena [17], M. S. Chapman in Atlanta [18], G. Rempe in Garching [19], and D. Meschede in Bonn [20] have capitalized on this combination and developed experimental setups based on optical cavities and atoms with low-lying electronic states. In this chapter, we shall describe in details the basic constituents of setup developed in the group of D. Meschede.

4.1 Optical cavity

An optical cavity is the heart of setup that we exploit in this manuscript. This cavity represents an open resonator that consists of two polished spherical mirrors facing each other and where each mirror has a diameter of about 1 mm and a radius of curvature $R_M = 5$ cm. The cavity supports resonant frequency $\omega \approx 2\pi \cdot 0.35$ PHz that almost matches the $F = 4 \to F' = 5$ (D$_2$-line) transition of a Cs atom (see below). Moreover, the mirrors are separated by the distance $L = 159$ μm (in the origin of $x - y$ plane) and they accommodate about $k = 373$ antinodes along the cavity axis such that $L \approx k\,\lambda/2$, and where $\lambda = 2\pi c/\omega \approx 852$ nm is the wavelength associated with the resonant cavity mode. By using special coating and polishing techniques of the cavity mirrors, moreover, the cavity relaxation time $1/\kappa \simeq 0.3$ μs has been achieved in the group of D. Meschede [79].

As for the microwave cavity, the transverse cavity field components are described by the cavity field structure (1.27), where the cavity waist w is determined by the curvature R_M, distance L, and resonant wavelength λ [see (3.1)]. In the group of D. Meschede, these parameters have been chosen such that the cavity waist is equal to about 23 μm. The optical cavity, in addition, as well supports two linearly and orthogonally polarized modes of light (C_x and C_y) with polarizations located in the $x - y$ plan and which are separated by a birefringent splitting of about $\delta = 2\pi \cdot 200$ KHz. Moreover, the cavity mode volume is calculated by using the expression $V = \pi w^2 L/4 \approx 6.6 \cdot 10^4$ μm^3 and the amplitude of vacuum field in the cavity center [$f(\mathbf{r}') = 1$] is given by the expression (3.2) and yields the value $\mathcal{E} \approx 432$ Volt/m.

We explained in the previous chapter that in order to realize the atom-cavity energy exchange, the mean number of residual thermal photons should be negligible. In contrast to the microwave cavity, however, the optical cavity is insensitive to the thermal photons even at the room temperature. It can be readily checked by substituting the cavity frequency ω and $T \approx 300$ K in the expression (3.3), that the average number of thermal photons satisfies the condition $n_{\text{th}} \ll 1$ which ensures that the contribution of residual thermal field can be safely neglected. Optical cavities, therefore, can be utilized without any special cooling mechanism, which is one notable advantage if compared with the

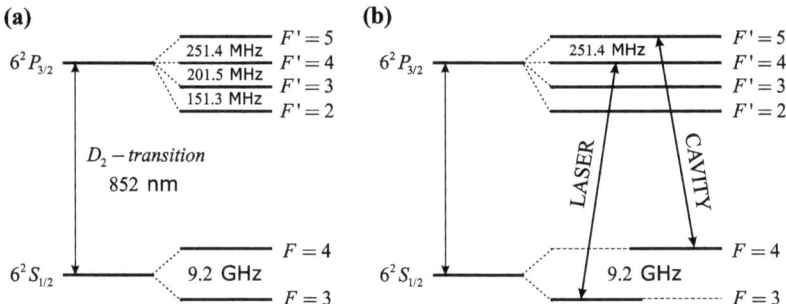

Figure 4.1: (a) Level structure of the first excited state in the atom of Cs [96]. (b) The same level structure interpreted as a three-level Λ-type configuration.

microwave cavities.

Experiments in which one or more atoms couple to the cavity, require one precise control of cavity resonance frequency with respect to a given atomic transition frequency. This control, in turn, requires an excellent mechanical stability of the resonator since thermal drifts and mechanical vibrations inevitably lead to uncertainties in the distance between the mirrors. The passive mechanical stabilization (as utilized for microwave cavities) is not sufficient in the case of an optical cavity since the optical wavelengths are much shorter and an active stabilization is then needed to compensate these uncertainties. This stabilization is usually based on the Pound-Drever-Hall feedback scheme [92], in which the cavity resonant frequency is compared with a stable laser source and a piezoelectric stack (placed under the cavity mirror) is then used to compensate for uncertainties in order to match the cavity resonant frequency with the frequency of stable laser light. By using such an active stabilization scheme, the uncertainties in the distance between cavity mirrors have been reduced to about $0,45 \cdot 10^{-12}$ m in the group of D. Meschede.

4.2 Neutral atoms as qubits

As we mentioned in section 3.2, the atoms represent flying qubits which need to be strongly coupled to the cavity field while they pass through the resonator. In contrast to the case of Rydberg atoms, however, the atoms with low-lying electronic states possess rather moderate dipole momentum which, nevertheless, is compensated by a small volume of an optical cavity. The level structure of first excited state in the atom of Cs is displayed in Fig. 4.1(a). As we mentioned in the previous section, the cavity supports

CHAPTER 4: Optical cavity setup

resonant frequency that almost matches the $F = 4 \to F' = 5$ transition of a *Cs* atom. The decay rate associated to this transition is $\gamma/2\pi = 2.6$ MHz and the respective radiative lifetime is $1/\gamma \approx 61$ ns. The same atomic transition, moreover, is characterized by the (circular) transition polarization ϵ_a^+ and by the dipole matrix element $3.17\,q\,a_0$ (compare to $1767\,q\,a_0$ for Rydberg atom with $n = 50$).

As seen from Fig. 4.1(a), furthermore, an atom of *Cs* has two hyperfine ground levels with $F = 3$ and $F = 4$. These two levels together with the first excited state can be interpreted as the three-level Λ-type configuration displayed in Fig. 4.1(b) and which is required for our (entangled state generation) scheme from the second chapter. In this configuration, moreover, the cavity mode is coupled off-resonantly to the $F = 4 \to F' = 5$ transition, while the laser beam to the $F = 3 \to F' = 4$ transition. In contrast to a three-level Λ-type atom in which the transition between the two lower states is forbidden, the transition $F = 3 \to F = 4$ can be realized by means of a microwave field. However, since cavity and laser fields cannot realize this transition alone, the atom of *Cs* can be safely considered as a three-level atom in the Λ-type configuration [79].

In contrast to the Rydberg atoms, moreover, the atoms with low-lying electronic states do not require any special preparation sequence and could be simply effused from an oven. However, since the effused atoms need to be coupled to the cavity mode in a well controllable fashion, additional tools are required to store a predetermined amount of atoms and transport them coherently inside the cavity. The two basic tools which provide the necessary degree of control over atoms are (i) the mageto-otical trap (MOT) that plays the role of a deterministic source of atoms and (ii) the optical lattice (conveyor belt) that transports atoms into the cavity from the MOT along with an position control over the atomic motion. These two tools being combined in the same setup, therefore, enable one to initiate and store atoms in the MOT and then to insert them in the sites of an optical lattice and transport inside the cavity [16, 18, 20]. The detailed description of a MOT can be found in Ref. [93], while a brief description of a conveyor belt is given below.

4.2.1 Transportation of atoms

In order to control the position of atoms, an optical lattice (conveyor belt) has been integrated into the experimental setup in the group of D. Meschede. A conveyor belt is a far detuned dipole trap in which atoms are attracted to regions of high laser intensity. The corresponding force arises from the interaction of the induced dipole moment with the gradient of light field [97]. As displayed schematically in Fig. 4.2, the trap consists of two counter-propagating Gaussian beams of Nd:YAG laser (1064 nm, 4 W) which create together a standing wave interference pattern with periodical potential wells of

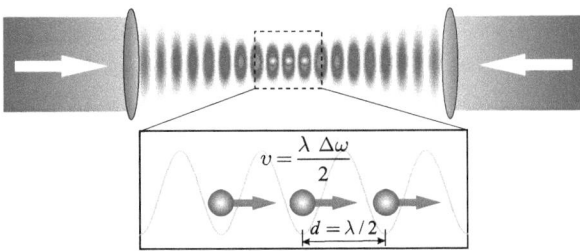

Figure 4.2: Schematic view of atoms in an optical lattice (conveyor belt). Two focussed and counter-propagating laser beams with frequencies $\omega_1 = \omega + \Delta\omega/2$ and $\omega_2 = \omega - \Delta\omega/2$ give rise to an interference pattern in the field intensity with a series of equidistant potential wells in which the atoms can be trapped. The distance between two neighbored wells is given by (half of) the lattice wavelength λ, while the velocity of belt is determined by the detuning $\Delta\omega$ of two laser beams.

532 nm separation. On the one end, the lattice is overlapped with the MOT such that the atoms can be efficiently inserted into the belt, and on the other end, the lattice is sandwiched by the cavity mirrors. The number-triggered insertion procedure enables to insert a predetermined amount of atoms in the lattice [98], while the use of an additional optical lattice permits to rearrange the atoms in order to make them equally distanced from each other [99].

By changing the frequencies of laser beams by means of two acousto-optic modulators, the interference pattern is set into motion and carries the atoms along the lattice axis. This conveyor belt, therefore, enables to transport atoms over macroscopic distances (up to 10 mm) with a sub-micrometer precision [100]. Since the optical excitation in the optical lattice can be kept very low, a conveyor belt provides a nearly conservative trapping potential that is especially important for experiments on quantum information. It has been experimentally demonstrated that an optical lattice preserves the coherence of transported atoms and can be utilized as a holder of a quantum register for storing quantum information. By encoding the quantum information in the hyperfine levels ($F = 3$ and $F = 4$), moreover, the storage time of about 6 s has been reported in Refs. [101, 102].

4.3 Summary

The three-level atoms inserted in the sites of a conveyor belt and a detuned optical cavity are the main ingredients of our scheme for generation of W entangled states which we

CHAPTER 4: Optical cavity setup

shall present in the third part of this manuscript. We mentioned in the previous chapter that in order to exploit in practice the entanglement of atoms that is based on the off-resonant cavity, each atom from the chain should be strongly coupled to the cavity, i.e., both the conditions $g_\circ \gg \kappa$ and $g_\circ \gg \gamma$ must be satisfied at the same time. In this chapter, we mentioned that the rates associated with atomic decay and cavity relaxation are $\kappa/2\pi = 0.4$ MHz and $\gamma/2\pi = 2.6$ MHz, respectively. At the same time, the vacuum Rabi frequency at the cavity center (where the cavity field amplitude takes its maximum) is about $g_\circ/2\pi = 10$ MHz and it corresponds to a Rabi period $2\pi/g_\circ = 0.1$ μs. We can conclude, therefore, that the mentioned conditions for the strong coupling regime are nicely satisfied and this is possible merely due to small cavity volume and high positioning precision of atoms with respect to the cavity antinode.

We also mention that a series of remarkable experiments has been performed and reported in the group of G. Rempe. For instance, the photon-photon entanglement with a single atom coupled to an optical cavity [104], cavity based cooling of single atoms [105], cavity based control of a single photon's polarization, and phase [106, 107], and e.t.c. It is also obvious that any practical realization of theoretical schemes based on the cavity(-laser) mediated interaction of atoms suffers from various practical limitations: (i) imperfect Rabi pulses, (ii) cavity relaxation and atomic decay, (iii) imperfect positioning precision of atoms with respect to the cavity antinode, to name just a few of them. Various improvements, therefore, are further needed in order to overcome these difficulties and achieve an excellent control of the cavity mediated entanglement of atoms in the framework of optical cavity-QED.

Part III

Multipartite entangled states for chains of atoms

Chapter 5

Generation of entangled states with a microwave cavity

In the first part of this manuscript we introduced the resonant interaction regime between an (circularly polarized) two-level atom and a cavity which supports two orthogonally polarized modes of light field. We found that such an atom interacts only with one cavity mode and that its interaction with another (orthogonally polarized) cavity mode can be neglected if the birefringent splitting is sufficiently large with respect to the atom-cavity coupling. We showed, moreover, that by setting an appropriate interaction time and an initial atom-cavity state, we can control the coherent energy exchange of coupled atom-cavity system and generate an entangled atom-cavity state. We concluded, therefore, that the resonant atom-cavity interaction is an exceptional regime that can be used to entangle one single two-level atom (atomic qubit) with the photon field of a cavity (cavity qubit) in a well-controlled way.

In the the second part of this manuscript, furthermore, we described an experimental setup developed in the Laboratoire Kastler Brossel by S. Haroche and co-workers [34, 35]. This setup is schematically displayed in Fig. 5.1(a) and it includes (i) one microwave cavity that supports two orthogonally polarized modes of light field, (ii) several Ramsey plates, and (iii) a chain of rubidium atoms prepared in highly excited Rydberg states such that the atomic transition between two neighbor levels can be tuned in resonance with one or another cavity mode (see below). Moreover, atoms of rubidium are prepared in one of the three Rydberg levels with principal quantum numbers 51, 50, or 49, and which are denoted as excited state $|e\rangle$, ground state $|g\rangle$, and auxiliary state $|a\rangle$, respectively. Owing to the particular design of microwave cavity, however, only the states $|e\rangle$ and $|g\rangle$ can be involved in the atom-cavity interaction because only the $e \leftrightarrow g$ transition frequency of a rubidium atom can be tuned to the frequencies of cavity modes. The classical microwave field injected between the Ramsey plates by the sources S and S_d, in contrast, can be

CHAPTER 5: Generation of entangled states with a microwave cavity

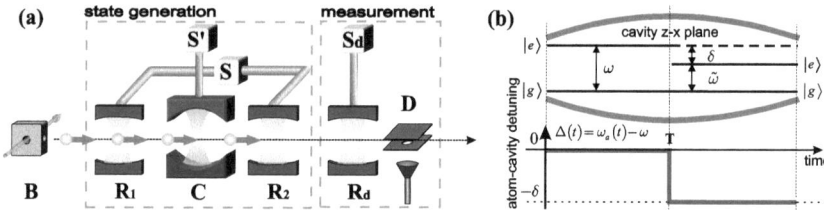

Figure 5.1: (a) Schematic setup of an experiment in which a chain of Rydberg atoms is emitted from a source B and then passes through a Ramsey zone R_1, a cavity C, the Ramsey zones R_2 and R_d, until the atoms are ionized one-by-one in the detector D. The classical fields in the Ramsey zones are generated by the microwave sources S, S' and S_d. (b) Temporal matching of the $e \leftrightarrow g$ atomic transition frequency ω_a to either the frequency ω of cavity mode C_x or the frequency $\widetilde{\omega}$ of mode C_y which is produced in the course of atom-cavity interaction. The lower part of this figure displays the (time-dependent) step-wise change of atom-cavity detuning $\Delta(t) = \omega_a(t) - \omega$ such that for $t < T$ the atom is resonant with the mode C_x and for $t > T$ with the mode C_y. See text for further discussions.

adapted to drive the $e \leftrightarrow g$ or $g \leftrightarrow a$ transitions and is utilized to manipulate the superposition between these atomic states as we explained in section 3.2.1.

An entangled state of a Rydberg atom (A) with the photon field of cavity is achieved in a deterministic way by tuning the $e \leftrightarrow g$ transition frequency $\omega_a(t)$ as a function of time such that the atom is in resonance with either cavity mode C_x or C_y while it passes through the cavity. For a sufficiently fast switch of the detuning $\Delta(t) = \omega_a(t) - \omega$, the resonant interaction is realized with either mode C_x for $\Delta(t < T) = 0$ or with the mode C_y for $\Delta(t > T) = -\delta$ as displayed in Fig. 5.1(b), and where a step-wise change from the $A - C_x$ to the $A - C_y$ interaction is assumed. In the experiments by S. Haroche and coworkers, for instance, the detuning is changed by applying a well adjusted time-varying electric field across the gap between the cavity mirrors, so that the required (Stark) shift of atomic $e \leftrightarrow g$ transition frequency is obtained [42]. However, an atom can interact resonantly only with one of cavity modes since the second mode is then frozen from the interaction due to a (large) birefringent splitting $\delta = \omega - \widetilde{\omega}$. The overall $A - C_x - C_y$ time evolution of atom-cavity state, therefore, can be safely separated into two independent parts (i) the evolution that occurs due to the $A - C_x$ resonant interaction as given by Eqs. (1.74) and (ii) the evolution due to $A - C_y$ resonant interaction as given by Eqs. (1.75). In the latter evolution, moreover, the imaginary factor arises due to orthogonal polarization of mode C_x with respect to mode C_y.

5.1. Entangled states with a single-mode cavity

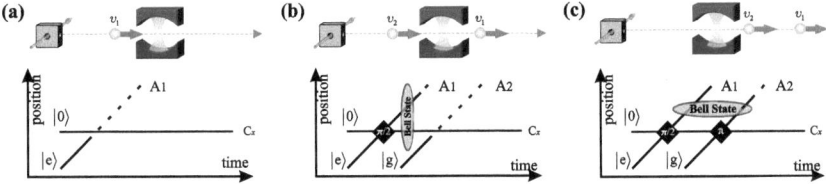

Figure 5.2: Three snapshots of the sequence that generates the entangled state (5.1) between atoms A_1 and A_2. The upper part of each sub-figure indicates how the atoms pass through the cavity, while the lower part shows the corresponding temporal sequence with the trajectories of atoms and cavity. The pictograms used in these figures are described in the text.

The experimental setup displayed in Fig. 5.1(a), therefore, provides all the necessary ingredients to generate complex entangled states of cavity photon field with Rydberg atoms which pass sequentially through the cavity. By exploiting one (or more) such resonant cavities, in this chapter, we shall propose and discuss the schemes which enable one to (i) generate multi-partite GHZ and W states, (ii) generate one- and two-dimensional cluster states of arbitrary size, and (iii) prove the entanglement formation of three- and four-partite Greenberger-Horne-Zeilinger (GHZ) and W states.

5.1 Entangled states with a single-mode cavity

As a prerequisite for the next sections, we shall work out three schemes to generate the multipartite Greenberger-Horne-Zeilinger (GHZ), W, and linear cluster states, in which one single cavity mode is utilized [60, 61]. In other words, we confine ourselves to the situation in which the atoms are tuned in resonance with one single cavity mode, say C_x, while passing sequentially through the cavity. Our purpose is to provide the individual interaction times between the cavity mode and each atom which are required to generate a particular entangled state for a chain of Rydberg atoms that is initially prepared in the product state. Moreover, we shall introduce a convenient graphic language in order to display these steps in terms of temporal sequences and quantum circuits.

Before we turn to GHZ and W states, however, let us consider the most simple case of an atomic chain that consists of two atoms being initially prepared in the product state $|e_1, g_2\rangle$. In this case, we shall display the necessary interaction times for which these two atoms produce the entangled state

$$|\Psi^{\text{Bell}}_{A_1-A_2}\rangle = \frac{1}{\sqrt{2}} \left(|e_1, g_2\rangle - |g_1, e_2\rangle \right) \tag{5.1}$$

CHAPTER 5: Generation of entangled states with a microwave cavity

after both passing through the cavity. For this, assume that the cavity mode C_x is initially empty and the first atom is emitted (by the atomic source) in the excited state with the velocity $v_1 = 2\,g_\circ\,w/\sqrt{\pi}$ which corresponds to the atom-cavity interaction time $t_1 = \pi/(2\,g_\circ)$. According to the atom-cavity evolution (1.74a) and definition (1.60), the entangled Bell state

$$|\Psi_{A_1-C_x}^{\text{Bell}}\rangle = \frac{1}{\sqrt{2}}(|e_1;\,0\rangle + |g_1;\,1\rangle) \qquad (5.2)$$

is generated after the first atom (A_1) leaves the cavity.

In order to generate the state (5.1) from (5.2), moreover, we need to map the cavity state upon the second atom. This is done by sending the second atom in the ground state with the velocity $v_2 = g_\circ\,w/\sqrt{\pi}$ which corresponds to the atom-cavity interaction time $t_2 = \pi/g_\circ$. By using Eqs. (1.74) it can be readily checked that the cavity states $\{|0\rangle, |1\rangle\}$ are mapped upon the atomic states $\{|g_2\rangle, |e_2\rangle\}$, respectively, while the cavity mode is factored out in the vacuum state. In other words, the cavity state has been mapped to the atomic state and, therefore, the desired state (5.1) is produced from (5.2) (for further details, see Ref. [38] where this two-step sequence has been demonstrated experimentally). Notice that the rotation angles $g_\circ\,t_1 = \pi/2$ and $g_\circ\,t_2 = \pi$ determine completely the evolution of coupled atom-cavity systems $A_1 - C_x$ and $A_2 - C_x$, respectively. Moreover, these two angles (so-called Rabi pulses: $p_1 = \pi/2$ and $p_2 = \pi$) along with definition (1.60) determine the velocities v_1 and v_2 which are necessary for atoms A_1 and A_2 in order to realize the required atom-cavity evolution.

In Figure 5.2, furthermore, by using the graphical language introduced by S. Haroche and co-workers, we displayed three snapshots of above scheme with all the manipulations which atoms A_1 and A_2 undergo while crossing the cavity. Specifically, the lower part of each sub-figure shows the temporal diagram in which the black diamond indicates the evolution of coupled atom-cavity system along with respective rotation angle (Rabi pulse) displayed inside. With this simple but illustrative example, therefore, we are ready to discuss more elaborated schemes for generation of multipartite entangled states with a chain of Rydberg atoms.

5.1.1 W states

In this section, we discuss the generation of W state [22]

$$|\Psi_N^W\rangle = \frac{1}{\sqrt{N}}(e^{i\,\psi}\overbrace{|\downarrow_1,\uparrow_2,\ldots,\uparrow_N\rangle + |\uparrow_1,\downarrow_2,\ldots,\uparrow_N\rangle + \ldots + |\uparrow_1,\uparrow_2,\ldots,\downarrow_N\rangle}^{N\text{ terms}}) \qquad (5.3)$$

for a chain of N atoms prepared initially in the product state $|e_1, g_2, \ldots, g_N\rangle$ and where we consider the correspondence $\{|\uparrow_i\rangle = |g_i\rangle, |\downarrow_i\rangle = |e_i\rangle\}$ between the qubit states and a pair of neighbor levels of i-th atom in the chain. For this state, a sequence of Rabi

5.1. Entangled states with a single-mode cavity

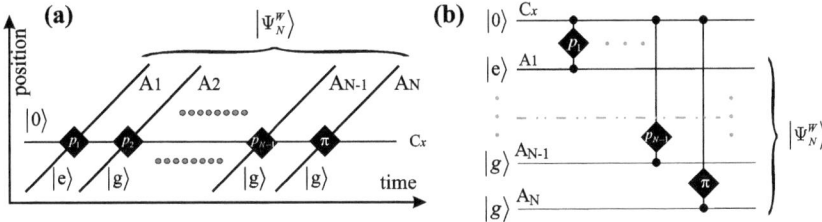

Figure 5.3: (a) Temporal sequence for generation of the W state associated with the chain of atoms $A_1, \ldots, A_N$. (b) The corresponding quantum circuit in which the cavity is represented by the uppermost line, while the atoms by the lines below. The pictograms in these figures are described in the text.

pulses can be worked out and expressed as a temporal sequence for the passage of atoms through the cavity. This sequence is displayed in Fig. 5.3(a) with the individual Rabi pulses (atom-cavity rotations) given by [60]

$$p_n = g_\circ t_n = \begin{cases} 2 \arccos\left(\frac{1}{\sqrt{N}}\right), & n = 1; \\ 2 \arccos\left(\sqrt{\frac{N-n}{N-n+1}}\right), & n = 2, \ldots, N-1. \end{cases} \quad (5.4)$$

This temporal sequence looks similarly to those from Fig. 5.2, however, it includes N atomic trajectories and also contains different Rabi pulses associated with individual atom-cavity interaction. By using the Eqs. (1.74), moreover, it can be shown that (up to a constant phase) the state (5.3) with $\psi = \pi$ is generated from the initial product state $|e_1, g_2, \ldots, g_N\rangle$, and where the cavity mode is factored out in the vacuum state.

Besides of displaying the individual interactions between the atoms and cavity, that is the particular sequence of Rabi pulses, the representation of unitary transformation as a quantum circuit is shown in Fig. 5.3(b). Obviously, both representations (a) and (b) are equivalent and can be utilized on purpose, and where the latter one can be translated into the language of quantum gates [4].

5.1.2 GHZ states

In this section, we discuss the generation of Greenberger Horne Zeilinger (GHZ) state [36]

$$|\Psi_N^{GHZ}\rangle = \frac{1}{\sqrt{2}} \left(e^{i\psi}|\uparrow_1, \ldots, \uparrow_N\rangle + |\downarrow_1, \ldots, \downarrow_N\rangle\right), \quad (5.5)$$

for a chain of N atoms prepared initially in the product state $|e_1, g_2, \ldots, g_N\rangle$, and where the qubit states $\{|\uparrow_i\rangle, |\downarrow_i\rangle\}$ refer to a pair of neighbor states $\{|g\rangle, |e\rangle\}$ or $\{|a\rangle, |g\rangle\}$ of i-th atom in the chain. In addition to the interaction with the cavity photon field, this

CHAPTER 5: Generation of entangled states with a microwave cavity

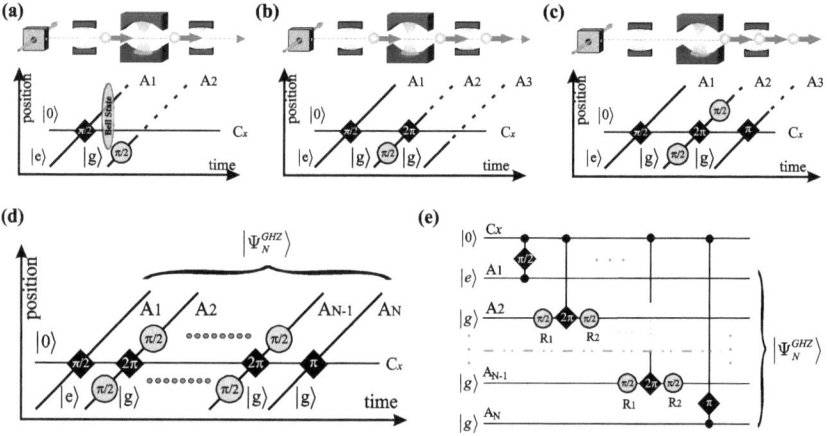

Figure 5.4: (a)-(c) Three snapshots of the sequence that generates the GHZ state (5.7) between the atoms A_1, A_2, and A_3. The upper part of each sub-figure indicates how the atoms pass through the cavity and Ramsey plates, while the lower part shows the corresponding temporal sequence with the trajectories of atoms and cavity. (d) Temporal sequence for generation of the GHZ state (5.9) associated with the chain of atoms $A_1, \ldots, A_N$. (b) The corresponding quantum circuit in which the cavity is represented by the uppermost line, while the atoms by the lines below. The pictograms in these figures are described in the text.

scheme requires the interaction of Rydberg atoms with a classical microwave field that is injected by the field source S between the Ramsey plates located in front and behind the cavity [see Fig. 5.1(a)]. We showed in section 3.2.1, moreover, that a microwave field source can drive $e \leftrightarrow g$ or $g \leftrightarrow a$ atomic transition and it produces the rotation (3.9) expressed in the basis $\{|g\rangle, |e\rangle\}$ or $\{|a\rangle, |g\rangle\}$, respectively. Below, we refer to this rotation as the Rasmey pulse $\hat{R}(\phi, \varphi)$ and denote it in our figures by grey circles indicating the rotation angle ϕ and, if needed, also the value of phase φ. In addition, we shall supply the subscripts R_1 or R_2 to these circles in order to associate such pulses to the Ramsey plates located in front or behind the cavity.

Before we turn to multipartite GHZ state, however, let us consider the most simple case of an atomic chain that consists of three atoms prepared initially in the product state $|e_1, g_2, g_3\rangle$. To generate a GHZ state for this chain, we first generate the atom-cavity entangled state (5.2) by sending the first atom (A_1) through the cavity with velocity $2 g_\circ w/\sqrt{\pi}$ that corresponds to a $\pi/2$ Rabi pulse. Next to A_1, we send the second atom (A_2) with velocity $g_\circ w/(2\sqrt{\pi})$ that corresponds to a 2π Rabi pulse. Just before A_2

enters the cavity, its state is transformed into the superposition $|g_2\rangle \to \frac{1}{\sqrt{2}}(|a_2\rangle + |g_2\rangle)$ by using a $\hat{R}_1(\pi/2, 0)$ Ramsey pulse tuned to the $g \leftrightarrow a$ transition frequency [see (3.9) for $\varphi = 0$], while A_2 crosses the plates R_1 in front of cavity. After the second atom has left the Ramsey plates R_1, it enters the cavity and interacts with the mode C_x for a 2π Rabi pulse. The effect of this Rabi pulse can be seen from Eqs. (1.74) which imply the evolution $|e_2; 0\rangle \to -|e_2; 0\rangle$ and $|g_2; 1\rangle \to -|g_2; 1\rangle$. At this point, therefore, the composite $A_1 - A_2 - C_x$ state becomes $[|e_1, (a_2 + g_2); 0\rangle + |g_1, (a_2 - g_2); 1\rangle]/2$.

After the atom A_2 has passed through the cavity, it is subjected again to a $\hat{R}_2(\pi/2, 0)$ pulse inside the second pair of Ramsey plates (located behind the cavity), thus, leading to the entangled GHZ state for two atoms and a cavity mode

$$|\Psi^{GHZ}_{A_1-A_2-C}\rangle = \frac{1}{\sqrt{2}}(|e_1, a_2; 0\rangle - |g_1, g_2; 1\rangle). \tag{5.6}$$

In order to generate the GHZ state for a chain of three atoms, furthermore, we need to map the cavity state upon the atom A_3. This mapping is performed by sending the third atom prepared in the ground state through the cavity such that $A_3 - C_x$ system undergoes a π Rabi pulse. It can be checked that after A_3 leaves the cavity, the composite state (5.6) becomes

$$|\Psi^{GHZ}_3\rangle = \frac{1}{\sqrt{2}}(|e_1, a_2, g_3\rangle + |g_1, g_2, e_3\rangle), \tag{5.7}$$

where the cavity state is factored out in the vacuum state. Obviously, the generated state (5.7) is equivalent to the GHZ state (5.5) for $N = 3$ and $\psi = 0$ under the change of notation

$$\{|\uparrow_1\rangle = |e_1\rangle, |\downarrow_1\rangle = |g_1\rangle\}, \quad \{|\uparrow_2\rangle = |a_2\rangle, |\downarrow_2\rangle = |g_2\rangle\}, \quad \{|\uparrow_3\rangle = |g_3\rangle, |\downarrow_3\rangle = |e_3\rangle\}. \tag{5.8}$$

In Figs. 5.4(a)-(c), we displayed three snapshots of above steps with all the manipulations which atoms A_1, A_2, and A_2 undergo while crossing the cavity and Ramsey plates. A more detailed discussion of these manipulations is given in Ref. [41], where this sequence of Ramsey and Rabi pulses was demonstrated experimentally. At this point, moreover, we are ready to introduce our scheme for generation of multipartite GHZ state (5.5) for a chain of N atoms prepared initially in the product state $|e_1, g_2, \ldots, g_N\rangle$. The respective temporal sequence and the quantum circuit are displayed in Figs. 5.4(d) and (e), respectively. By using the Eqs. (1.74) and (3.9), furthermore, it can be checked that this sequence results into the state

$$|\Psi^{GHZ}_N\rangle = \frac{1}{\sqrt{2}}\left(e^{i\pi(N+1)}|e_1, a_2, \ldots, a_{N-1}, g_N\rangle + |g_1, g_2, \ldots, g_{N-1}, e_N\rangle\right) \tag{5.9}$$

and where the cavity mode is factored out in the vacuum state. Obviously, the generated state (5.9) is equivalent to the GHZ state (5.5) under the change of notation

$$\{|\uparrow_1\rangle = |e_1\rangle, |\downarrow_1\rangle = |g_1\rangle\}, \quad \{|\uparrow_i\rangle = |a_i\rangle, |\downarrow_i\rangle = |g_i\rangle\}, \quad \{|\uparrow_N\rangle = |g_N\rangle, |\downarrow_N\rangle = |e_N\rangle\} \tag{5.10}$$

CHAPTER 5: Generation of entangled states with a microwave cavity

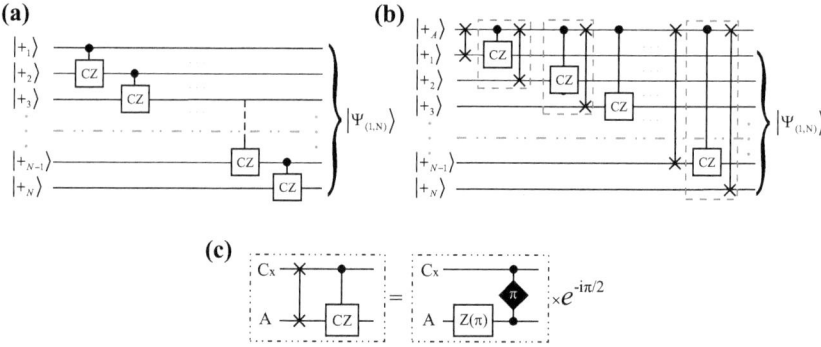

Figure 5.5: (a) Quantum circuit for generation of a linear cluster state between N uncorrelated qubits. Each qubit is initially prepared in the $|+\rangle = (|\uparrow\rangle + |\downarrow\rangle)/\sqrt{2}$ state, and one controlled-z gate is applied subsequently to any two neighboring qubits. (b) Alternative quantum circuit for linear cluster state generation, in which the controlled-z operation is successively applied to the ancilla qubit A and k-th qubit, and followed by the swapping of ancilla state with the $k+1$-th qubit. (c) Two equivalent circuits that follow from relation (5.14) after multiplying it from the right side by $[\hat{Z}(\pi) \otimes \hat{I}]^{-1}$.

with $i = 2, \ldots, N-1$, and where the phase $\psi = \pi(N+1)$ depends on the total number of atoms in the chain.

5.1.3 Linear cluster state

As we mentioned in the introduction, H. J. Briegel and R. Raussendorf have introduced a novel type of multi-partite entangled states in Ref. [48]. These (so-called) cluster states are known to exhibit a rather high persistency and robustness of their entanglement with regard to decoherence effects [49]. Apart from the fundamental interest in these states [50] and their use in quantum communication protocols [51], the cluster states are the key ingredient for one-way quantum computations [52].

In this section, we discuss the generation scheme of a linear $1 \times N$ cluster state for a chain of atoms, in which only one cavity mode C_x is utilized. This scheme was first suggested by C. Schön and co-autors [47] and is adapted here for the cavity setup as displayed in Fig. 5.1(a) [61]. The linear cluster state is defined as [48]

$$|\Psi_{(1,N)}\rangle = \frac{1}{2^{N/2}} \bigotimes_{i=1}^{N} \left(|\uparrow_i\rangle + |\downarrow_i\rangle \Upsilon_{i+1}\right), \qquad (5.11)$$

where $\Upsilon_k = |\uparrow_k\rangle\langle\uparrow_k| - |\downarrow_k\rangle\langle\downarrow_k|$ acts on the k-th qubit such that $\Upsilon_{N+1} \equiv 1$, and

5.1. Entangled states with a single-mode cavity

where we consider the assignment $\{|\uparrow_i\rangle = |g_i\rangle, |\downarrow_i\rangle = |e_i\rangle\}$ between the qubit states and a pair of neighbor levels of i-th atom in the chain. Equivalently, one can define the linear cluster state as a one-dimensional lattice of N qubits, where the nodes refer to the qubits initialized in the product state $|+_1\rangle \times \ldots \times |+_N\rangle$ with $|+\rangle = (|\uparrow\rangle + |\downarrow\rangle)/\sqrt{2}$, and where the edges of lattice refer to the two-qubit controlled-z gate [4]

$$|\uparrow_1\rangle|\uparrow_2\rangle \to |\uparrow_1\rangle|\uparrow_2\rangle, \quad |\uparrow_1\rangle|\downarrow_2\rangle \to |\uparrow_1\rangle|\downarrow_2\rangle, \quad |\downarrow_1\rangle|\uparrow_2\rangle \to |\downarrow_1\rangle|\uparrow_2\rangle, \quad |\downarrow_1\rangle|\downarrow_2\rangle \to -|\downarrow_1\rangle|\downarrow_2\rangle$$

which is applied between all the neighboring nodes. According to this latter definition, Fig. 5.5(a) displays the successive interactions which are necessary to generate the linear cluster state for N initially uncorrelated qubits. Instead of applying the controlled-z gate to each pair of neighboring qubits k and $k+1$, however, we can apply this two-qubit gate to the ancilla qubit and ordinary qubit k, and then swap the state of ancilla qubit with qubit $k+1$ as displayed in Fig. 5.5(b). Note that in this circuit, we have inserted one additional swap gate between the ancilla and the first atom which has no effect on the output cluster state since the ancilla qubit is prepared initially in the state $|+\rangle$.

Below we shall associate the ancilla qubit with the cavity mode C_x and the ordinary qubits with the Rydberg atoms. According to the second scheme from Fig. 5.5(b), this identification implies that the atoms pass sequentially through the cavity and that only one atom couples to the cavity at a time, which fits nicely to our cavity setup that we consider in this chapter. As seen from Fig. 5.5(b), moreover, only two types of unitary gates have to be realized between the cavity mode and each atom which passes through the cavity, namely, (i) a swap gate followed by the controlled-z gate for atoms $A_1 \ldots A_{N-1}$ and (ii) a swap gate for the atom A_N. Before the atom-cavity interaction takes place, moreover, each atom must be prepared in the superposition $|+\rangle = (|e\rangle + |g\rangle)/\sqrt{2}$. In the setup displayed in Fig. 5.1(a), this superposition is achieved by preparing the atoms in the excited state and then performing the rotation

$$|e\rangle \to \frac{1}{\sqrt{2}}(|e\rangle + |g\rangle) \tag{5.12}$$

just before the atom enters the cavity. As explained above, the rotation (5.12) is realized efficiently by using a $\hat{R}_1(\pi/2, 0)$ Ramsey pulse tuned to the $e \leftrightarrow g$ transition frequency [see (2.53a) for $\varphi = 0$] while the atom crosses the Ramsey plates in front of cavity.

It can be checked, furthermore, that the atom-cavity evolution that corresponds to a π Rabi pulse is equivalent to the modified swap gate [see (1.74)]

$$\hat{U}^{\text{m-swap}} = \begin{pmatrix} 1 & 0 & 0 & 0 \\ 0 & 0 & -1 & 0 \\ 0 & 1 & 0 & 0 \\ 0 & 0 & 0 & 1 \end{pmatrix}, \tag{5.13}$$

CHAPTER 5: Generation of entangled states with a microwave cavity

expressed in the basis $\{|g;0\rangle, |g;1\rangle, |e;0\rangle, |e;1\rangle\}$. In contrast to the conventional swap gate (which has no minus sign), below we shall refer to this two-qubit operation as to the m-swap gate. Following the work by Schön and co-authors [47], moreover, we express the m-swap gate (5.13) in the form

$$\hat{U}^{\text{m-swap}} = (-i)\, \hat{U}^{\text{cz}} \cdot \hat{U}^{\text{swap}} \cdot \left(\hat{Z}(\pi) \otimes \hat{I}\right), \tag{5.14}$$

where

$$\hat{U}^{\text{swap}} = \begin{pmatrix} 1 & 0 & 0 & 0 \\ 0 & 0 & 1 & 0 \\ 0 & 1 & 0 & 0 \\ 0 & 0 & 0 & 1 \end{pmatrix}, \quad \hat{U}^{\text{cz}} = \begin{pmatrix} 1 & 0 & 0 & 0 \\ 0 & 1 & 0 & 0 \\ 0 & 0 & 1 & 0 \\ 0 & 0 & 0 & -1 \end{pmatrix} \tag{5.15}$$

are the swap and controlled-z gates taken in the same basis as the matrix (5.13), and where $\hat{Z}(\xi) \equiv e^{-i\hat{\sigma}_z \xi/2}$ denotes the atomic rotation operator. Thus, the equality (5.14) implies that the m-swap gate is equivalent (up to a constant phase) to a swap gate followed by a controlled-z gate together with a local rotation of atomic state. In order to realize only the swap gate followed by the controlled-z gate as required by our scheme [see Fig. 5.5(b)], therefore, the m-swap gate (atom-cavity π rotation) should be preceded by the local rotation $\hat{Z}^{-1}(\pi) = -\hat{Z}(\pi)$ of atomic state as displayed in Fig. 5.5(c).

Up to this point, we just summarized a scheme that enables one to generate a linear $1 \times N$ cluster state (5.11) by sending a chain of N uncorrelated atoms through the cavity in such a way that only one atom couples to the cavity mode at a time. Specifically, we have shown that each atom is incorporated into the cluster state by performing the superposition (5.12) followed by one more $\hat{Z}(\pi)$ rotation of atomic state and finalized by a a π Rabi pulse. In order to fully adapt this scheme for our cavity setup, it is necessary to express the $\hat{Z}(\pi)$ atomic rotations in terms of Ramsey pulses which can be generated by the microwave source S. By using the unitary matrix (3.9), it can be readily checked that

$$\hat{Z}(\pi) = \hat{R}(\pi, \pi/2) \cdot \hat{R}(\pi, 0) \tag{5.16}$$

can be realized by applying two π Ramsey pulses successively, such that the first pulse is resonant and the second pulse is detuned by $\pi/2$. In section 3.2.1, we explained how a sequence of two short (off-resonant) pulses realize the rotation $\hat{R}(\phi, \varphi) \cdot \hat{R}(\phi, 0)$ with $\phi = \Omega_\circ \tau$ and $\varphi = (\omega_a - \omega_J) T$. We also explained that the time delay T is controlled by switching on and off the source of microwave field while the atom passes through the Ramsey plates. We conclude, therefore, that the rotation (5.16) can be realized efficiently by means of Ramsey plates R_1 and, therefore, we can express the equality (5.14) in the form

$$\begin{aligned} \hat{U}^{\text{cz}} \cdot \hat{U}^{\text{swap}} &= (-i)\, \hat{U}^{\text{m-swap}} \cdot \left(\hat{Z}(\pi) \otimes \hat{I}\right) \\ &= (-i)\, \hat{U}^{\text{m-swap}} \cdot \left(\left[\hat{R}_1(\pi, \pi/2) \cdot \hat{R}_1(\pi, 0)\right] \otimes \hat{I}\right). \end{aligned} \tag{5.17}$$

5.2. Proving the entanglement generation

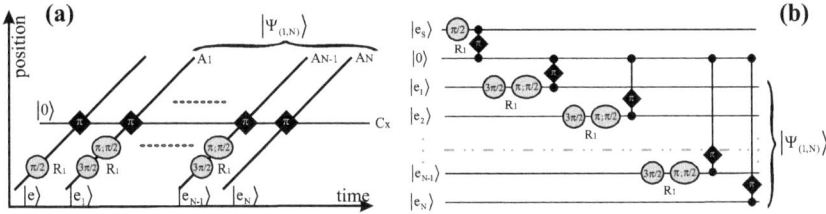

Figure 5.6: (a) Temporal sequence for generation of a linear cluster state that is encoded into a chain of N Rydberg atoms passing through the cavity. (b) Quantum circuit that corresponds to the above sequence. The pictograms and notation in these figures are explained in the text.

With this analysis, we now have all ingredients available to generate linear cluster states using our cavity setup shown in Fig. 5.1(a). The temporal sequence and the equivalent quantum circuit for this scheme are displayed in Fig. 5.6(a) and (b), respectively [61]. As before, the atom-cavity interactions are depicted by black diamonds and the Ramsey pulses $\hat{R}(\phi, \varphi)$ are shown as gray circles for which we indicate the interaction time in units of Rabi pulses ϕ and the phase φ (if it is non-zero). Note that, in order to prepare the cavity mode in the $|+\rangle$ state, we utilize an auxiliary atom A_s that is initialized in the excited state and which crosses the empty cavity before the chain of atoms arrives. This auxiliary atom interacts for a $\pi/2$ Ramsey pulse with the microwave field R_1 and then for a π Rabi pulse with the cavity. According to Eq. (5.12) and Eqs. (1.74), the initially empty cavity field is then set to the $|+\rangle = (|0\rangle + |1\rangle)/\sqrt{2}$ state, while the auxiliary atom is factored out in its ground state. Let us also note that the last swap gate between the cavity and the N-th atom is replaced by the m-swap gate (5.13), which simply maps the cavity state $|0\rangle$ upon the atomic ground state $|g\rangle$ and the cavity state $|1\rangle$ upon the excited state $|e\rangle$. This m-swap operation, which is the π Rabi pulse, finally factors out the cavity state from the atomic cluster state.

5.2 Proving the entanglement generation

Obviously, each scheme for generation of a particular entangled state (for a given atomic chain) should come along with a recipe that enables one to prove that the requested state has been indeed generated. Up to now, however, we were concerned with the (Rabi and Ramsey) pulses that are necessary to generate the desired entangled state and not much was said about the end point of atomic trajectories, namely, the detector D [see Fig. 5.1(a)]. To project the state of a Rydberg atom upon one of its levels e, g, or a,

CHAPTER 5: Generation of entangled states with a microwave cavity

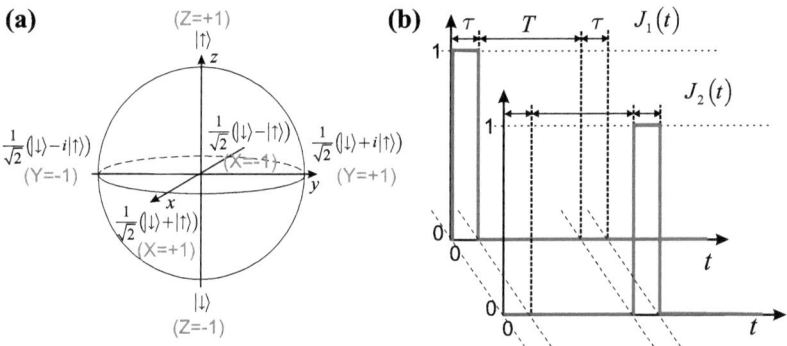

Figure 5.7: (a) Bloch sphere. See the text for description. (b) Temporal sequence of two short (off-resonant) Ramsey pulses $J_1(t)$ and $J_2(t)$ applied to the atoms A_1 and A_2 while the atoms pass through the Ramsey plates R_d. See text for explanation.

a field ionization technique is applied in the experiments by S. Haroche and co-workers [84]. From the signals measured by this detector (for many realizations of the same generation sequence), the probabilities $P(e_i)$, $P(g_i)$, and $P(a_i)$ which correspond to the electronic occupation of levels e, g, or a, respectively, are found for the i-th atom. This type of projective measurement is often referred in the literature as the *longitudinal* measurement (experiment).

To better understand why one distinct projective measurement – the *transversal* measurement need to be carried out, let us consider the GHZ state (5.7). With the probability 1/2, one expects to find the atomic chain either in the state $|e_1, a_2, g_4\rangle$ or $|g_1, g_2, e_3\rangle$ after the chain has left the cavity. However, the same probabilities are obtained also for the (uncorrelated) statistical mixture of corresponding basis states, for instance the mixed state $[\{1/2, |e_1, a_2, g_4\rangle\}, \{1/2, |g_1, g_2, e_3\rangle\}]$. Therefore, the longitudinal measurement taken alone is not sufficient for proving the non-classical nature of (quantum mechanically) correlated atoms and need to be augmented by additional measurements. The same line of reasoning applies to the Bell state $\frac{1}{\sqrt{2}}(|\downarrow_1,\uparrow_2\rangle + |\uparrow_1,\downarrow_2\rangle)$ that describes a rotation-invariant spin singlet state. For such a singlet state, moreover, one finds the two spins always in opposite direction for any choice of the quantization axis of (projective) measurement. In the literature, this counter-intuitive result is known also as Einstein-Rosen-Podolsky (EPR) paradoxon [5], and this freedom in the choice of quantization axis will be exploited in the next sections to reveal and display the non-classical correlations of generated entangled states.

In order to introduce a more quantitative description for the projective measurement

5.2. Proving the entanglement generation

in the framework of cavity-QED, let us introduce the geometrical language based on the Bloch sphere. Recall that an arbitrary state of a qubit can be parametrized by means of only two angles ϕ and φ

$$|\Psi\rangle = \cos(\phi/2)|\uparrow\rangle + \sin(\phi/2)\, e^{i\varphi}|\downarrow\rangle \qquad (5.18)$$

with the ranges $0 \leq \phi \leq \pi$ and $0 \leq \varphi \leq 2\pi$, respectively. These two angles, moreover, can be interpreted as the polar angles in three-dimensional spherical coordinates and lead to the representation of above state as a point on a unit sphere – the Bloch sphere. In the notation of (5.18), the states $|\uparrow\rangle$ and $|\downarrow\rangle$ are located on the z axis and coincide with the North and South poles of sphere as displayed in Fig. 5.7(a). These states $|\uparrow\rangle$ and $|\downarrow\rangle$, moreover, are the two eigenstates of spin operator $\hat{\sigma}_z$ with eigenvalues $Z = +1$ and $Z = -1$, respectively. In a similar fashion, therefore, the x axis is defined by the eigenstates $|+^x\rangle = \frac{1}{\sqrt{2}}(|\downarrow\rangle + |\uparrow\rangle)$ and $|-^x\rangle = \frac{1}{\sqrt{2}}(|\downarrow\rangle - |\uparrow\rangle)$ of spin operator $\hat{\sigma}_x$ with eigenvalues $X = +1$ and $X = -1$, respectively, and the y axis is defined by the eigenstates $|+^y\rangle = \frac{1}{\sqrt{2}}(|\downarrow\rangle + i|\uparrow\rangle)$ and $|-^x\rangle = \frac{1}{\sqrt{2}}(|\downarrow\rangle - i|\uparrow\rangle)$ of spin operator $\hat{\sigma}_y$ with eigenvalues $Y = +1$ and $Y = -1$, respectively [see Fig. 5.7(a)]. Given any state on the sphere, therefore, the diametrically opposite point will always represent orthogonal states. Any other axis $\xi(\varphi)$ in the equatorial $x - y$ plane, furthermore, corresponds to superpositions of $|\uparrow\rangle$ and $|\downarrow\rangle$ with equal weights ($\phi = \pi/2$) and is determined by the azimuthal angle φ. The axis $\xi(\varphi)$, therefore, is characterized (up to a global phase facrtor) by the states $|+^\varphi\rangle = \hat{Z}(-\varphi)|+^x\rangle = \frac{1}{\sqrt{2}}(|\downarrow\rangle + e^{i\varphi}|\uparrow\rangle)$ and $|-^\varphi\rangle = \hat{Z}(-\varphi)|-^x\rangle = \frac{1}{\sqrt{2}}(|\downarrow\rangle - e^{i\varphi}|\uparrow\rangle)$ where, again, the '+' and '−' signs are chosen to distinguish between positive and negative directions along the axes. Recall that the basis states $|\uparrow\rangle$ and $|\downarrow\rangle$ of i-th qubit are related to the two neighbor atomic states $\{|e\rangle, |g\rangle\}$ or $\{|g\rangle, |a\rangle\}$ via expressions $\{|g_i\rangle = |\uparrow_i\rangle, |e_i\rangle = |\downarrow_i\rangle\}$ for the N-partite W state (5.3) and expressions (5.10) for the N-partite GHZ state (5.5). Notice, moreover, that we defined the Bloch sphere such that the poles of sphere are located on the z axis and coincide with the (longitudinal) projection measurement performed by the detector.

Following Hagley and co-workers [38], we explain how the same detector (D) that is used for projection of atomic states along the (longitudinal) z axis, can be applied to perform a (transversal) projection along either the x or $\xi(\varphi)$ axes of Bloch sphere. Namely, we like to show that (i) a combination of $\hat{R}(\pi/2, 0)$ Ramsey pulse followed by the longitudinal measurement is equivalent to the transversal measurement upon the x axis of sphere and (ii) a combination of $\hat{R}(\pi/2, \varphi)$ Ramsey pulse followed by the longitudinal measurement is equivalent to the transversal measurement upon the $\xi(\varphi)$ axis of Bloch sphere. In other words, we state the following identities

$$|\langle -^x|\Psi\rangle|^2 = |\langle e|\hat{R}(\pi/2,0)|\Psi\rangle|^2, \quad |\langle +^x|\Psi\rangle|^2 = |\langle g|\hat{R}(\pi/2,0)|\Psi\rangle|^2, \qquad (5.19a)$$

$$|\langle -^\varphi|\Psi\rangle|^2 = |\langle e|\hat{R}(\pi/2,\varphi)|\Psi\rangle|^2, \quad |\langle +^\varphi|\Psi\rangle|^2 = |\langle g|\hat{R}(\pi/2,\varphi)|\Psi\rangle|^2, \qquad (5.19b)$$

CHAPTER 5: Generation of entangled states with a microwave cavity

where $|\Psi\rangle$ is an arbitrary state (5.18) and where the assignment $\{|g\rangle = |\uparrow\rangle, |e\rangle = |\downarrow\rangle\}$ has been considered. The validity of these identities follows from the identities

$$\hat{R}^\dagger(\pi/2,0)|e\rangle = \frac{1}{\sqrt{2}}(|e\rangle - |g\rangle), \quad \hat{R}^\dagger(\pi/2,0)|g\rangle = \frac{1}{\sqrt{2}}(|e\rangle + |g\rangle),$$

$$\hat{R}^\dagger(\pi/2,\varphi)|e\rangle = \frac{1}{\sqrt{2}}(|e\rangle - e^{i\varphi}|g\rangle), \quad \hat{R}^\dagger(\pi/2,\varphi)|g\rangle = \frac{e^{-i\varphi}}{\sqrt{2}}(|e\rangle + e^{i\varphi}|g\rangle).$$

The identities (5.19), therefore, allow us to identify a pair or Ramsey plates together with the ionization detector as a type of detector that projects the state of a Rydberg atom upon the (transversal) quantization axes x or $\xi(\varphi)$, and where the assignment $\{|a\rangle = |\uparrow\rangle, |g\rangle = |\downarrow\rangle\}$ should be considered if the microwave field injected between the Ramsey plates is tuned to the $a \leftrightarrow g$ atomic transition.

In order to perform such transversal measurements, therefore, we have inserted one extra pair of Ramsey plates R_d (along with microwave source S_d) before the detector in our setup [see Fig. 5.1(a)] and this enables to perform both longitudinal and transversal measurements by utilizing the microwave source S_d and tuning it to the atomic transition in question. Having discussed the Bloch sphere and the experimental realization of various measurements, in the next section, we shall describe the scheme to reveal the non-classical correlations associated with the most simplest entangled state – the Bell state (5.1).

5.2.1 Entanglement measure for an atomic Bell state

As we mentioned above, the Bell state (5.1) represents, in the spin language, a rotation-invariant spin singlet. For such a singlet state, moreover, one finds the two spins always pointing in the opposite direction for any choice of the quantization axis of (projective) measurement. In order to reveal such non-classical correlation between these spins, therefore, we project the state of first atom along the x axis and the state of second atom along the $\xi(\varphi)$ axis of Bloch sphere. In other word, we project the atomic states along two axes in the equatorial $(x - y)$ plane of Bloch sphere such that the angle φ becomes a parameter that is set by the off-resonant microwave pulse applied in R_d before the atoms are detected.

By using the identities (5.19), these projections are implemented by means of the sequence displayed in Fig. 5.8(a). This sequence can be divided in two parts: (i) $\hat{R}_d(\pi/2,0)$ acting upon A_1 followed by D, and (ii) $\hat{R}_d(\pi/2,\varphi)$ acting upon A_2 followed by D. According to this figure, moreover, the first Ramsey pulse pulse on atom A_1 is followed, after a time delay T, by a second Ramsey pulse on atom A_2 and each atomic trajectory is finalized by a projective measurement in the longitudinal basis. During the last Ramsey pulse, the phase difference $\varphi = \Delta_J T$ is accumulated, where the time delay T plays the

5.2. Proving the entanglement generation

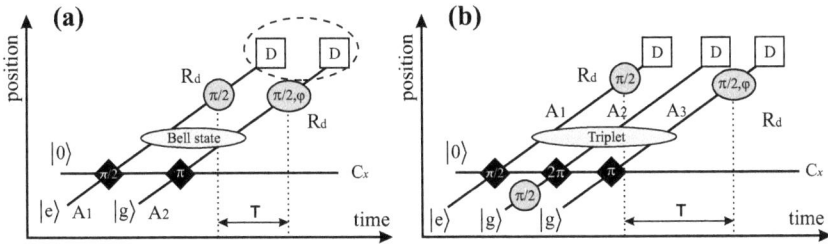

Figure 5.8: (a) Temporal sequence displaying the generation and transversal measurements for the Bell state (5.1) associated with atoms A_1 and A_2. (b) Temporal sequence displaying the generation and transversal measurements for the three-partite GHZ state (5.25) associated with atoms A_1, A_2, and A_3. See the text for explanations.

role of an adjustable parameter. Since each atom can be detected in one of two states, the sequence displayed in Fig. 5.8(a) yields four detection probabilities

$$P(e_1, e_2; \varphi) = |(\langle -\tfrac{x}{1}| \times \langle -\tfrac{\varphi}{2}|) |\Psi^{\text{Bell}}_{A_1-A_2}\rangle|^2, \qquad P(g_1, g_2; \varphi) = |(\langle +\tfrac{x}{1}| \times \langle +\tfrac{\varphi}{2}|) |\Psi^{\text{Bell}}_{A_1-A_2}\rangle|^2, \quad (5.21a)$$

$$P(e_1, g_2; \varphi) = |(\langle -\tfrac{x}{1}| \times \langle +\tfrac{\varphi}{2}|) |\Psi^{\text{Bell}}_{A_1-A_2}\rangle|^2, \qquad P(g_1, e_2; \varphi) = |(\langle +\tfrac{x}{1}| \times \langle -\tfrac{\varphi}{2}|) |\Psi^{\text{Bell}}_{A_1-A_2}\rangle|^2, \quad (5.21b)$$

which depend on the angle $\varphi \propto T$. These probabilities are combined for many repetitions of one and the same experiment in order to produce the correlation signal [68]

$$I(\varphi) = P(e_1, e_2; \varphi) + P(g_1, g_2; \varphi) - P(e_1, g_2; \varphi) - P(g_1, e_2; \varphi). \tag{5.22}$$

In section 3.2.1 we explained how a sequence of two short off-resonant pulses (displayed in Fig. 3.1) realizes the rotation $\hat{R}(\phi, 0) \cdot \hat{R}(\phi, \varphi)$ with $\phi = \Omega_\circ \tau$ and $\varphi = (\omega_a - \omega_J) T$. According to the sequence (of transversal measurements) displayed in Fig. 5.8(a), however, the rotations $\hat{R}_d(\pi/2, 0)$ and $\hat{R}_d(\pi/2, \varphi)$ are applied to different atoms which pass sequentially through the Ramsey plates and are separated by a time delay T. In order to show that a sequence of two short (off-resonant) pulses displayed in Fig. 5.8(a) leads to the set of probabilities (5.21), notice that this sequence corresponds to the pulse diagram displayed in Fig. 5.7(b), in which the functions $J_1(t)$ and $J_2(t)$ are associated to the atoms A_1 and A_2, respectively. In this case, the atom-field Hamiltonian corresponding to the interaction of each atom with the microwave field becomes

$$\hat{H}^s_i = -i\hbar \frac{\Omega_\circ}{2} \left(\hat{\sigma}_i^\dagger e^{i\Delta_J t} - \hat{\sigma}_i e^{-i\Delta_J t} \right) J_i(t); \quad i = 1, 2, \tag{5.23}$$

and which for the interaction time τ satisfying the condition $\Delta_J \tau \ll 1$, yields the evolutions

$$\exp\left[-\frac{\Omega_\circ}{2} \left(\hat{\sigma}_1^\dagger - \hat{\sigma}_1 \right) \tau \right], \quad \exp\left[-\frac{\Omega_\circ}{2} \left(\hat{\sigma}_2^\dagger e^{-i\Delta_J T} - \hat{\sigma}_2 e^{i\Delta_J T} \right) \tau \right]$$

CHAPTER 5: Generation of entangled states with a microwave cavity

associated to the atoms A_1 and A_2, respectively. These evolution operators, moreover, coincide with the rotation operators $\hat{R}(\Omega_\circ \tau, 0)$ and $\hat{R}(\Omega_\circ \tau, \Delta_J T)$, respectively, and must be applied to the Bell state (5.1) which is produced just after both atoms leave the cavity [see Fig. 5.8(a)]. It can be readily checked, furthermore, that after both atoms leave the Ramsey plates R_d, the composite atomic state becomes ($\varphi = \Delta_J T$)

$$\begin{aligned}
\hat{R}_d^{A_2}(\pi/2, \varphi)\, \hat{R}_d^{A_1}(\pi/2, 0)|\Psi_{A_1-A_2}^{\text{Bell}}\rangle &= \frac{1}{2} \hat{R}_d^{A_2}(\pi/2, \varphi)\left[(|g_1\rangle + |e_1\rangle)|g_2\rangle - (|g_1\rangle - |e_1\rangle)|e_2\rangle\right] \\
&= \frac{1}{2\sqrt{2}}\left[\left(1 - e^{-i\,\varphi}\right)|e_1, e_2\rangle + \left(1 - e^{i\,\varphi}\right)|g_1, g_2\rangle \right. \\
&\quad \left. + \left(1 + e^{i\,\varphi}\right)|e_1, g_2\rangle - \left(1 + e^{-i\,\varphi}\right)|g_1, e_2\rangle\right] \quad (5.24)
\end{aligned}$$

and which gives rise to the set of probabilities (5.21), if projected in the basis $|e_1, e_2\rangle$, $|g_1, g_2\rangle$, $|e_1, g_2\rangle$, and $|g_1, e_2\rangle$.

For an idealized experiment, the correlation signal (5.22) takes the form $I(\varphi) = -\cos(\varphi)$. This modulation can be explained by the following qualitative arguments. When $\varphi = 0$, both detections are performed along the x axis and the two atoms are found in opposite states, which implies $I(0) = -1$. When $\varphi = \pi$, the two detections are performed along opposite directions and the two atoms are found in the same state, which implies $I(\pi) = 1$. In the intermediate situation, when $\varphi = \pi/2$, the detection axes are orthogonal and there is no correlation, which implies $I(\pi/2) = 0$. The observed oscillations of signal (5.22) as function of φ, therefore, ensure that the generated atomic state is a spin singlet state and provides an entanglement measure for an arbitrary two-partite atomic state. This recipe, moreover, meets the mentioned request for carrying out an additional (transversal) measurement and reveals the non-classical nature of (quantum mechanically) correlated atoms prepared in the state (5.1).

5.2.2 Three-partite entangled GHZ and W states

In the previous section, we introduced an entanglement measure associated to a two-partite atomic state. This measure is based on the correlation signal (5.22) that is obtained after multiple realizations of one and the same experimental sequence. Although this recipe can be realized only within two atoms in the chain, in this section we show that the same technique enables to reveal quantum correlations of the three-partite GHZ state

$$|\Psi_3^{GHZ}\rangle = \frac{1}{2}\left[|e_1, (a_2 + g_2), g_3\rangle - |g_1, (a_2 - g_2), e_3\rangle\right], \quad (5.25)$$

which can be generated by using the sequence displayed in Figs. 5.4(a)-(c), with only difference that the second atom does not interact with the microwave field in R_2 after it leaves the cavity.

Our idea is to perform the transversal measurement on atoms A_1 and A_3 which are separated by a time delay T as displayed in Fig. 5.8(b), and where A_2 simply pass

5.2. Proving the entanglement generation

through the Ramsey plates R_d without interaction. It can be readily checked that after all the atoms leave the Ramsey plates, the composite atomic state becomes

$$\begin{aligned}\hat{R}_d^{A_3}(\pi/2,\varphi)\,\hat{R}_d^{A_1}(\pi/2,0)|\Psi_3^{GHZ}\rangle &= \frac{1}{4}\Big([\big(1-e^{-i\,\varphi}\big)|g_2,e_3\rangle + \big(1-e^{i\,\varphi}\big)|a_2,g_3\rangle \\ &\quad + \big(1+e^{i\,\varphi}\big)|g_2,g_3\rangle - \big(1+e^{-i\,\varphi}\big)|a_2,e_3\rangle]\,|g_1\rangle \\ &\quad + [\big(1-e^{-i\,\varphi}\big)|a_2,e_3\rangle + \big(1-e^{i\,\varphi}\big)|g_2,g_3\rangle \\ &\quad + \big(1+e^{i\,\varphi}\big)|a_2,g_3\rangle - \big(1+e^{-i\,\varphi}\big)|g_2,e_3\rangle]\,|e_1\rangle\Big),\end{aligned}$$

where $\varphi = \Delta_J T$. Atom A_1 enters first the detector and is projected onto the states $|e_1\rangle$ or $|g_1\rangle$ which, in turn, makes the above state to collapse into one of two wave-functions

$$|\Psi_+\rangle = \frac{1}{2\sqrt{2}}\left[\big(1-e^{-i\,\varphi}\big)|g_2,e_3\rangle + \big(1-e^{i\,\varphi}\big)|a_2,g_3\rangle + \big(1+e^{i\,\varphi}\big)|g_2,g_3\rangle - \big(1+e^{-i\,\varphi}\big)|a_2,e_3\rangle\right],$$

$$|\Psi_-\rangle = \frac{1}{2\sqrt{2}}\left[\big(1-e^{-i\,\varphi}\big)|a_2,e_3\rangle + \big(1-e^{i\,\varphi}\big)|g_2,g_3\rangle + \big(1+e^{i\,\varphi}\big)|a_2,g_3\rangle - \big(1+e^{-i\,\varphi}\big)|g_2,e_3\rangle\right],$$

where $|\Psi_+\rangle$ corresponds to the atom A_1 detected in the state $|g_1\rangle$ and $|\Psi_-\rangle$ corresponds to the atom A_1 detected in the state $|e_1\rangle$. These two wave-functions, furthermore, give rise to the correlation signals

$$I_\pm(\varphi) = P_\pm(g_2,e_3;\varphi) + P_\pm(a_2,g_3;\varphi) - P_\pm(g_2,g_3;\varphi) - P_\pm(a_2,e_3;\varphi) = \pm\cos(\varphi), \quad (5.26)$$

where the '+' sign is associated with $|\Psi_+\rangle$ (atom A_1 detected in $|g_1\rangle$) and '−' sign is associated with $|\Psi_-\rangle$ (atom A_1 detected in $|e_1\rangle$). Apart from the modulation due to time delay $\varphi \propto T$, therefore, one additional parameter – the sign '$\pm$' appears in the signals (5.26) and this sign is correlated with the (detected) state of A_1. We can conclude that the two-partite entanglement measure which we introduced in the previous section enables to reveal quantum correlations of the three-partite GHZ state (5.25).

As we shall explain below, moreover, the same technique can be successfully utilized for a three-partite W state in order to reveal the non-classical correlations associated to this state. To proceed, notice that the W state (5.3) for $N = 3$ can be expressed in the form

$$\frac{1}{\sqrt{3}}[|g_1,e_2,g_3\rangle + |g_1,g_2,e_3\rangle - |e_1,g_2,g_3\rangle] = \frac{1}{\sqrt{3}}[|g_1\rangle(|e_2,g_3\rangle + |g_2,e_3\rangle) - |e_1\rangle|g_2,g_3\rangle]. \quad (5.27)$$

The right part of this equality implies that atoms A_2 and A_3 are found in the Bell state $(|e_2,g_3\rangle + |g_2,e_3\rangle)/\sqrt{2}$ or in the product state $|g_2,g_3\rangle$ depending on the state of A_1 after detection. Therefore, one can perform the transversal measurement of atoms $A_2 - A_3$ and record only the probabilities $P(g_1,g_2,g_3;\varphi)$, $P(g_1,e_2,e_3;\varphi)$, $P(g_1,e_2,g_3;\varphi)$, $P(g_1,g_2,e_3;\varphi)$ for which A_1 has been detected in the ground state. These probabilities, in turn, yield the signal

$$P(g_1,e_2,e_3;\varphi) + P(g_1,g_2,g_3;\varphi) - P(g_1,e_2,g_3;\varphi) - P(g_1,g_2,e_3;\varphi) = \cos(\varphi) \quad (5.28)$$

CHAPTER 5: Generation of entangled states with a microwave cavity

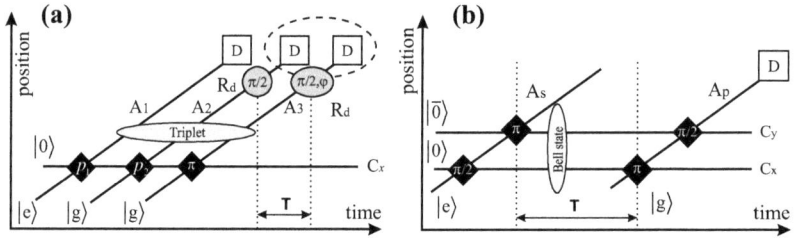

Figure 5.9: (a) Temporal sequence displaying the generation and transversal measurements for the three-partite W state (5.27) associated with atoms A_1, A_2, and A_3. The indicated Rabi pulses are $p_1 = 2\arccos(1/\sqrt{3})$ and $p_2 = 2\arccos(1/\sqrt{2})$. (b) Temporal sequence displaying the generation and transversal measurements for the Bell state (5.34) associated with cavity modes $C_x - C_y$. See the text for explanations.

with $\varphi = \Delta_J T$, and where the time delay T has been introduced between the rotations of atoms A_2 and A_3 while they passed through R_d. Owning to this recipe, the corresponding temporal sequence is displayed in Fig. 5.9(a).

The three-partite W state (5.27), however, can be expressed in the two equivalent forms

$$\frac{1}{\sqrt{3}}\left[|g_2\rangle\left(|g_1,e_3\rangle - |e_1,g_3\rangle\right) + |g_1,e_2,g_3\rangle\right] = \frac{1}{\sqrt{3}}\left[|g_3\rangle\left(|g_1,e_2\rangle - |e_1,g_2\rangle\right) + |g_1,g_2,e_3\rangle\right], \quad (5.29)$$

which imply that the pairs $A_1 - A_3$ and $A_1 - A_2$ are found in the Bell states if atoms A_2 and A_3 have been detected in the ground state, respectively. In order to reveal the non-classical correlations associated to the W state (5.29), therefore, the measurements from Fig. 5.9(a) are not sufficient and should be supplied by the measurements of Bell states $(|g_1,e_3\rangle - |e_1,g_3\rangle)/\sqrt{2}$ and $(|g_1,e_2\rangle - |e_1,g_2\rangle)/\sqrt{2}$ for which A_2 and A_3 have been detected in the ground states, respectively.

In the second (transversal) experiment, therefore, the states of A_1 and A_3 should be rotated by Ramsey pulses $\hat{R}_d(\pi/2, 0)$ and $\hat{R}_d(\pi/2, \varphi)$, respectively, and the probabilities $P(g_1, g_2, g_3; \varphi)$, $P(e_1, g_2, e_3; \varphi)$, $P(e_1, g_2, g_3; \varphi)$, $P(g_1, g_2, e_3; \varphi)$ should be recorded. There probabilities, in turn, yield the correlation signal

$$P(g_1, g_2, g_3; \varphi) + P(e_1, g_2, e_3; \varphi) - P(e_1, g_2, g_3; \varphi) - P(g_1, g_2, e_3; \varphi) = \cos(\varphi) \quad (5.30)$$

with $\varphi = \Delta_J T$, and where the time delay T should be introduced between the two rotations. In a similar fashion, the states of A_1 and A_2 should be rotated and the probabilities $P(g_1, g_2, g_3; \varphi)$, $P(e_1, e_2, g_3; \varphi)$, $P(e_1, g_2, g_3; \varphi)$, $P(g_1, e_2, g_3; \varphi)$ should be recorded in the third experiment. There probabilities, as before, yield the correlation signal

$$P(g_1, g_2, g_3; \varphi) + P(e_1, e_2, g_3; \varphi) - P(e_1, g_2, g_3; \varphi) - P(g_1, e_2, g_3; \varphi) = \cos(\varphi), \quad (5.31)$$

with $\varphi = \Delta_J T$. We conclude, therefore, that once the three-partite W state (5.27) has been generated, the correlation signals (5.28), (5.30), and (5.31) obtained for all three experiments, should be reasonably close to the predicted expression $\cos(\Delta_J T)$. It is also obvious that one similar procedure can be applied to any N-partite W state and it simply requires more (state selective) transversal experiments in order to reveal the correlations associated to all Bell pairs contained in a given W state.

5.2.3 Entanglement measure for a cavity Bell state

Beside of varying the angle $\varphi = \Delta_J T$ and recording the modulation of correlation signal (5.22), there is another technique to perform an independent measurement on an arbitrary two-qubit state. In contrast to the previous scheme, however, this technique operates with the qubits encoded in the cavity modes C_x and C_y.

Recall that the birefringent splitting that we mentioned in section 1.2, produces the energy difference $\hbar\delta = \hbar(\omega - \widetilde{\omega})$ between the cavity (single-photon) states $|1\rangle$ and $|\bar{1}\rangle$ associated with modes C_x and C_y, respectively. The composite cavity states $|1,\bar{0}\rangle$ and $|0,\bar{1}\rangle$, therefore, correspond to the energies $E_{|1,\bar{0}\rangle} = \hbar\omega$ and $E_{|0,\bar{1}\rangle} = \hbar(\omega - \delta)$, if the vacuum energy is taken as reference. The freely evolving cavity wave-function that includes both composite states $|1,\bar{0}\rangle$ and $|0,\bar{1}\rangle$, hence, is governed by the Hamiltonian $\hat{H}_\delta = \hbar\omega|1,\bar{0}\rangle\langle1,\bar{0}| + \hbar(\omega - \delta)|0,\bar{1}\rangle\langle0,\bar{1}|$ and gives rise to the time-evolution

$$e^{-\frac{i}{\hbar}\hat{H}_\delta t} = e^{-i\omega t}\left(|1,\bar{0}\rangle\langle1,\bar{0}| + e^{i\delta t}|0,\bar{1}\rangle\langle0,\bar{1}|\right). \tag{5.32}$$

This evolution implies that after the time interval T, for instance between generation and collapse events, the freely evolving cavity wave-function is transformed according to the expression

$$\alpha|1,\bar{0}\rangle + \beta|0,\bar{1}\rangle \longrightarrow \alpha|1,\bar{0}\rangle + e^{i\delta T}\beta|0,\bar{1}\rangle, \tag{5.33}$$

where $|\alpha|^2 + |\beta|^2 = 1$ and where the overall phase $e^{-i\omega t}$ has been omitted by considering an appropriate interaction picture. We can conclude, therefore, that once the mode C_y is populated with one photon, the relative phase factor $e^{i\delta t}$ arises and is caused solely by the birefringent splitting between the (orthogonally polarized) cavity modes.

The technique we like to introduce in this section, reveals the non-classical correlations of a two-partite quantum state encoded in the cavity modes C_x and C_y. This technique is based on the Bell state

$$|\Psi^{\text{Bell}}_{C_x-C_y}\rangle = \frac{1}{\sqrt{2}}\left(i|0,\bar{1}\rangle + |1,\bar{0}\rangle\right) \tag{5.34}$$

that is generated by sending the source atom A_s prepared in the excited state through empty cavity such that A_s interacts first with mode C_x ($\Delta = 0$) for a $\pi/2$ Rabi pulse

CHAPTER 5: Generation of entangled states with a microwave cavity

and afterwards with mode C_y ($\Delta = -\delta$) for a π Rabi pulse as displayed in Fig. 5.9(b). Once the state (5.34) is generated, the cavity state evolves freely during the time delay T until its state is probed by one further atom A_p. We just explained that for a freely evolving cavity, the relative phase factor $e^{i\delta T}$ arises due to energy difference between the modes and implies the state $\left(i\, e^{i\delta T}\, |0,\bar{1}\rangle + |1,\bar{0}\rangle\right)/\sqrt{2}$ to be produced right before A_p couples to the cavity.

In the latter step, A_p prepared in the ground state interacts with mode C_x for a π Rabi pulse and afterwards with mode C_y for a $\pi/2$ Rabi pulse. According to Eqs. (1.74)-(1.75), therefore, the entire sequence displayed in Fig. 5.9(b) generates the wave-function

$$|\Psi_{A_p-C_y}\rangle = \frac{1}{2}\left[|e;\bar{0}\rangle\left(1+e^{i\delta T}\right) + i\,|g;\bar{1}\rangle\left(1-e^{i\delta T}\right)\right], \quad (5.35)$$

where the factored state $|0\rangle$ is not shown for brevity. After A_p leaves the cavity, its state is projected by the detector D and the probability to detect A_p in the exited or ground states

$$P(e;T) = \frac{1+\cos(\delta T)}{2}, \quad P(g;T) = \frac{1-\cos(\delta T)}{2}, \quad (5.36)$$

respectively, is recorded for a given time delay T. After multiple realizations of one and the same experimental sequence with a different time delays T, the observed oscillations of probabilities (5.36) ensure that the Bell state (5.34) has been indeed generated and the entire technique provides us with one additional entanglement measure similar to the Bell signal (5.22) introduced in the previous sections.

We point out that the idea behind this technique is to converted the phase δT, that arises solely due to free evolution of the entangled state (5.34), into the atomic amplitudes $\left(1 \pm e^{i\delta T}\right)$ which imply the modulation of probabilities (5.36) and which, in turn, can be observed by means of (repetitive) projective measurements. Further details concerning this technique can be found in Ref. [42], where both the generation sequence and measurements were demonstrated experimentally.

5.2.4 Four-partite entangled GHZ

In section 5.2.1, we used the signal (5.22) to reveal the non-classical correlations associated with the Bell state (5.1). The same technique, furthermore, has been adapted for the three-partite GHZ state (5.25) in section 5.2.2. It has been observed that the state of A_1 can be correlated with the sign of oscillations (5.26) which, in turn, ensure that the remaining two-partite wave-functions associated with atoms A_1 and A_2, are the Bell states. In the case of four-partite GHZ state, however, the signals (5.26) alone are not sufficient since there is no additional parameters which could be correlated with the state of fourth atom A_4.

5.2. Proving the entanglement generation

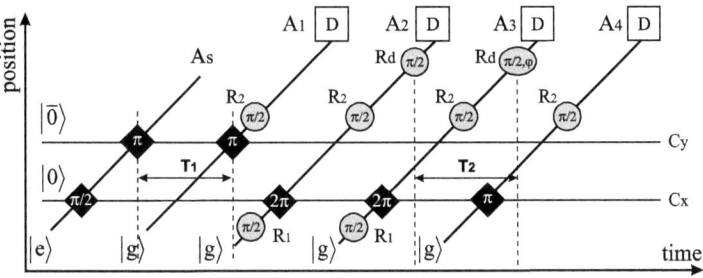

Figure 5.10: Temporal sequence displaying the generation and transversal measurements for state (5.37) which is associated with atoms A_1, A_2, A_3, and A_4. See the text for explanations.

In this section, we shall combine both techniques from sections 5.2.1 and 5.2.3 in order to reveal the non-classical correlations of four-partite GHZ state

$$|\Psi_4^{GHZ}\rangle = \frac{1}{2\sqrt{2}} \left(e^{i\vartheta} |(g_1+e_1), a_2, a_3, (g_4-e_4)\rangle + |(g_1-e_1), g_2, g_3, (g_4+e_4)\rangle \right), \quad (5.37)$$

which is generated for atoms $A_1 - A_4$ with the qubits being encoded in the atomic states $\{(|g_i\rangle + |e_i\rangle)/\sqrt{2}, (|g_i\rangle - |e_i\rangle)/\sqrt{2}\}$ ($i = 1, 4$) and $\{|a_i\rangle, |g_i\rangle\}$ ($i = 2, 3$). In order to produce this state, the Bell state (5.34) associated with the cavity modes $C_x - C_y$ is first generated. Once generated, this cavity state evolves freely such that the relative phase $\delta t = (\omega - \tilde{\omega}) t$ is produced. The grown of this phase is frozen at $t = T_1$ by the atom A_1 which maps the cavity state C_y. Next to A_1, two atoms A_2 and A_3 are entangled with the cavity mode C_x by using 2π Rabi pulses. Finally, the last atom A_4 maps the mode C_x and the state (5.37) is generated, where the (constant) phase is $\vartheta = \delta T_1$. The entire sequence is displayed in Fig. 5.10(a).

Atom A_1 exits first the Ramsey plates R_2 and is detected in the basis $\{|g_1\rangle, |e_1\rangle\}$. The measured state of A_1, however, is not relevant for transversal measurement since the phase $\vartheta = \delta T_1$ contains sufficient information about the coherent coupling of A_1 to the cavity mode C_x. Recall that the phase δt has been frozen at $t = T_1$ when the mode C_y has been mapped to the atom A_1 and, therefore, the correct parametrical dependence of output probability amplitudes (see below) on the time delay T_1 would ensure that the state (5.34) has been generated and that the state of C_y has been coherently mapped to A_1. Thus, we shall consider only the events in which A_1 is detected in the ground state $|g_1\rangle$ and shall discard all the events with $|e_1\rangle$. After detection of A_1, hence, the wave-function (5.37) collapses and the resulting state becomes

$$|\Psi_3^{GHZ}\rangle = \frac{1}{2} \left(e^{i\vartheta} |a_2, a_3, (g_4-e_4)\rangle + |g_2, g_3, (g_4+e_4)\rangle \right). \quad (5.38)$$

CHAPTER 5: Generation of entangled states with a microwave cavity

Next to A_1, atoms A_2, A_3, and A_4 pass through the Ramsey plates R_d such that A_2 and A_3 are separated by the time delay T_2. Similarly to section 5.2.2, we rotate the atomic states of A_2 and A_3 by means of $\hat{R}_d(\pi/2, 0)$ and $\hat{R}_d(\pi/2, \varphi)$ Ramsey pulses and the last atom A_4 simply pass through Ramsey plates R_d without interaction. It can be readily checked that after all the atoms leave the Ramsey plates, the composite atomic state becomes ($\vartheta = \delta T_1$, $\varphi = \Delta_J T_2$)

$$\hat{R}_d^{A_3}(\pi/2, \varphi) \hat{R}_d^{A_2}(\pi/2, 0) |\Psi_3^{GHZ}\rangle = \frac{1}{4} \Big(\Big[\Big(1 - e^{i(\vartheta - \varphi)}\Big) |a_2, g_3\rangle + \Big(e^{i\vartheta} + e^{i\varphi}\Big) |a_2, a_3\rangle$$
$$+ \Big(1 + e^{i(\vartheta - \varphi)}\Big) |g_2, g_3\rangle - \Big(e^{i\vartheta} - e^{i\varphi}\Big) |g_2, a_3\rangle \Big] |g_4\rangle$$
$$+ \Big[\Big(1 - e^{i(\vartheta - \varphi)}\Big) |g_2, g_3\rangle + \Big(e^{i\vartheta} + e^{i\varphi}\Big) |g_2, a_3\rangle$$
$$+ \Big(1 + e^{i(\vartheta - \varphi)}\Big) |a_2, g_3\rangle - \Big(e^{i\vartheta} - e^{i\varphi}\Big) |a_2, a_3\rangle \Big] |e_4\rangle \Big),$$

and gives rise to the set of probabilities

$$P_\pm(a_2, a_3; T_1, T_2) = P_\pm(g_2, g_3; T_1, T_2) = \frac{1 \pm \cos(\delta T_1 - \Delta_J T_2)}{4}, \tag{5.39}$$

$$P_\pm(a_2, g_3; T_1, T_2) = P_\pm(g_2, a_3; T_1, T_2) = \frac{1 \pm \cos(\delta T_1 - \Delta_J T_2)}{4}, \tag{5.40}$$

where the sign '+' corresponds to the $|g_4\rangle$ detection event and '−' to the detection event $|e_4\rangle$. As in previous section, these probabilities are combined for many realizations of the same temporal sequence and produce together the correlation signals

$$I_\pm(T_1, T_2) = P_\pm(a_2, a_3; T_1, T_2) + P_\pm(g_2, g_3; T_1, T_2) - P_\pm(a_2, g_3; T_1, T_2) - P_\pm(g_2, a_3; T_1, T_2)$$

which take the form $I_\pm(T_1, T_2) = \pm \cos(\delta T_1 - \Delta_J T_2)$, and where the time delays T_1 and T_2 are the two independent parameters which are manipulated separately.

The temporal sequence from Fig. 5.10(a) has to be realized for many times in order to reconstruct the signals $I_\pm(T_1, T_2)$ as functions of parameters T_1 and T_2. We remark that while T_1 is utilized to reveal the entanglement of Bell state (5.34), T_2 reveals the entanglement of remaining three-partite GHZ state (5.38). Moreover, the state selective measurements should be utilized in this scheme in order to collect only the probabilities for which A_1 has been detected in the ground state. If the four-partite GHZ state (5.37) has been generated in the first part of our scheme, then the signals $I_\pm(T_1, T_2)$ should be reasonably close to the predicted expressions: $\pm \cos(\delta T_1 - \Delta_J T_2)$ (see also [60]).

5.3 Two-dimensional cluster states

In contrast to the linear cluster states which we discussed in section 5.1.3, the two-dimensional cluster states enable to perform multi-qubit gate operations (e.g. quantum gates that act on two or more qubits simultaneously) in the framework of one-way computation [52]. These states, therefore, may provide a viable alternative to the conventional

5.3. Two-dimensional cluster states

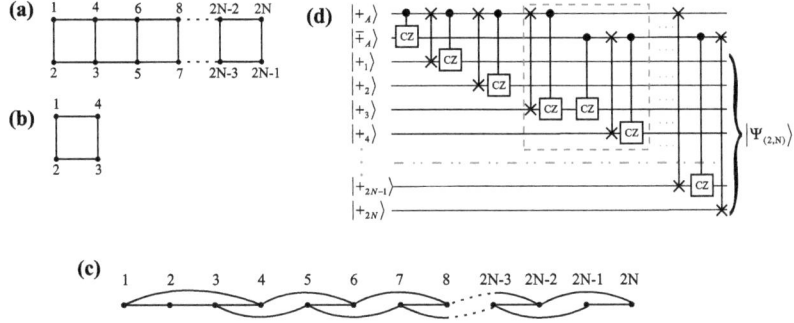

Figure 5.11: (a) Two-dimensional $2 \times N$ cluster state. (b) Box state (5.41) that is the simplest two-dimensional cluster state. (c) Definition of edges for a chain of $2N$ atoms (nodes), such that an effective two-dimensional $2 \times N$ cluster state is produced. The labels of nodes in Figure (a) correspond to the serial numbers of atoms inside the chain. (d) Quantum circuit for generation of the $|\Psi_{(2,N)}\rangle$ cluster state between $2N$ initially uncorrelated qubits, and for which two ancilla qubits are utilized. In this circuit, the controlled-z gates (edges) are applied according to sub-figure (c).

(circuit) computation in which sequences of unitary gates need to be implemented. Up to the present, however, only a minor progress has been achieved with regard to cavity-QED based schemes which generate two-dimensional cluster states [59]. Below, we suggest two practical schemes for the generation of $2 \times N$ and $3 \times N$ cluster states which are feasible for modern cavity-QED experiments. We describe in details the individual steps in the interaction of Rydberg atoms with the cavity modes C_x and C_y which are required to generate these states. Finally, we show how these schemes can be extended towards the generation of $M \times N$ two-dimensional cluster states of arbitrary size [61].

5.3.1 2×N cluster state

In this section, we introduce a scheme that generates two-dimensional cluster states $|\Psi_{(2,N)}\rangle$, i.e. states which form a two-dimensional lattice of $2 \times N$ qubits, where all the qubits (nodes of lattice) are initialized in the state $|+_1\rangle \times \ldots \times |+_{2N}\rangle$ and the controlled-z gate [4]

$$|\uparrow_1\rangle|\uparrow_2\rangle \to |\uparrow_1\rangle|\uparrow_2\rangle, \quad |\uparrow_1\rangle|\downarrow_2\rangle \to |\uparrow_1\rangle|\downarrow_2\rangle, \quad |\downarrow_1\rangle|\uparrow_2\rangle \to |\downarrow_1\rangle|\uparrow_2\rangle, \quad |\downarrow_1\rangle|\downarrow_2\rangle \to -|\downarrow_1\rangle|\downarrow_2\rangle$$

is applied for all edges that connect neighboring nodes as displayed in Fig. 5.11(a). We notice that a cluster state is neither restricted to a rectangular pattern of nodes not that only nearest neighbors could be connected with each other by means of a controlled-z

CHAPTER 5: Generation of entangled states with a microwave cavity

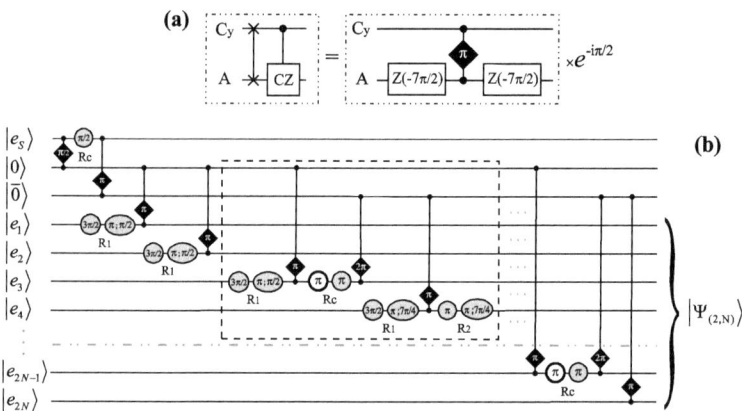

Figure 5.12: (a) Two equivalent circuits that follow from relation (5.43) after multiplying it from the right side by $[\hat{Z}(7\pi/2) \otimes \hat{Z}(7\pi/2)]^{-1}$ and where the imaginary factor has been omitted for brevity. (b) Quantum circuit for generation of the $2 \times N$ cluster state that is associated with a chain of $2N$ Rydberg atoms passing through the cavity. The new white-circled pictogram and R_c notation are explained in the text.

operation. However, we shall confine ourselves to the two-dimensional clusters with a rectangular pattern.

The simplest example of such a two-dimensional cluster state is the box state [53]

$$|\Psi_{(2,2)}\rangle = \frac{1}{2}(|\uparrow_1, +_2, \uparrow_3, +_4\rangle + |\uparrow_1, -_2, \downarrow_3, -_4\rangle + |\downarrow_1, -_2, \uparrow_3, -_4\rangle + |\downarrow_1, +_2, \downarrow_3, +_4\rangle), \quad (5.41)$$

where $|\pm\rangle = (|\uparrow\rangle \pm |\downarrow\rangle)/\sqrt{2}$. According to the setup displayed in Fig. 5.1(a), however, only a single chain of atoms is emitted by the atomic source and sent into the cavity. For this reason, we need to apply a different procedure (if compared with the linear cluster states) for defining the edges between the nodes associated with a chain of $2N$ atoms such that in the end, after all the atoms have crossed the cavity, an effective two-dimensional cluster state is generated. As we explained in the beginning of this chapter, our cavity supports two (orthogonally polarized) modes which can be addressed individually by the atoms and, therefore, these two modes can be considered as two independent photon qubits. This makes it possible to implement schemes in which two (photonic) ancilla qubits are involved and are associated with the modes C_x and C_y. This enables, in turn, the generation of $2N$-partite entangled state displayed in Fig. 5.11(c) and which represents the two-dimensional $2 \times N$ cluster state upon the assignment of atomic positions in a chain of $2N$ atoms to the two-dimensional cluster state as shown in Fig. 5.11(a). The quantum circuit that accomplishes this task is displayed in Fig. 5.11(d),

5.3. Two-dimensional cluster states

where the gates placed inside the dash-boxed area need to be repeated $N-3$ times [61]. Apart from the $A-C_x$ unitary gate which we introduced in section 5.1.3 [see Fig. 5.5(c)], this circuit contains three additional gates with two of them acting upon the $A-C_y$ system: (i) the swap gate followed by the controlled-z gate, and (ii) a single controlled-z gate. The third gate is the controlled-z gate that acts upon the C_x-C_y system.

Recall that Eqs. (1.75) describe the resonant evolution of an atom coupled to the cavity mode C_y ($\Delta = -\delta$), which for a π Rabi pulse yields the i-swap gate [65]

$$\hat{U}^{\text{i-swap}} = \begin{pmatrix} 1 & 0 & 0 & 0 \\ 0 & 0 & i & 0 \\ 0 & i & 0 & 0 \\ 0 & 0 & 0 & 1 \end{pmatrix}, \quad (5.42)$$

expressed in the basis $\{|g;\bar{0}\rangle, |g;\bar{1}\rangle, |e;\bar{0}\rangle, |e;\bar{1}\rangle\}$. Similar as in Eq. (5.14), we express this gate in the form

$$\hat{U}^{i-swap} = i \left(\hat{Z}(7\pi/2) \otimes \hat{I}\right) \cdot \hat{U}^{cz} \cdot \hat{U}^{swap} \cdot \left(\hat{Z}(7\pi/2) \otimes \hat{I}\right). \quad (5.43)$$

The swap gate followed by the controlled-z gate, therefore, is realized by a π Rabi pulse and two local $\hat{Z}^{-1}(7\pi/2) = \hat{Z}(-7\pi/2)$ rotations of the atomic state

$$\hat{U}^{cz} \cdot \hat{U}^{swap} = (-i) \left(\hat{Z}(-7\pi/2) \otimes \hat{I}\right) \cdot \hat{U}^{i-swap} \cdot \left(\hat{Z}(-7\pi/2) \otimes \hat{I}\right) \quad (5.44)$$

as displayed in Fig. 5.12(a). Each of atomic rotations, in turn, is efficiently realized by two off-resonant Ramsey pulses

$$\hat{Z}(-7\pi/2) = -\hat{R}(\pi, 7\pi/4) \cdot \hat{R}(\pi, 0) \quad (5.45)$$

which are applied while the atom passes through Ramsey plates R_1 (first z-rotation) and R_2 (second z-rotation), respectively.

In section 5.1.2, we already used the fact that a qubit can be encoded into the two neighboring states $|g\rangle$ nd $|a\rangle$ of a Rydberg atom. According to Eqs. (1.75), moreover, the atomic qubit encoded by means of states $\{|g\rangle, |a\rangle\}$ interacts with cavity mode C_y

$$|a;\bar{0}\rangle \to |a;\bar{0}\rangle, \quad |a;\bar{1}\rangle \to |a;\bar{1}\rangle, \quad |g;\bar{0}\rangle \to |g;\bar{0}\rangle, \quad |g;\bar{1}\rangle \to -|g;\bar{1}\rangle \quad (5.46)$$

for the case of a 2π Rabi pulse. Apparently, this transformation is the same as the controlled-z gate (5.15) if the states $\{|a;\bar{0}\rangle, |a;\bar{1}\rangle, |g;\bar{0}\rangle, |g;\bar{1}\rangle\}$ are taken as the basis. Therefore, we can use one 2π Rabi pulse to implement single $A-C_y$ controlled-z gate by carrying out the following three steps: If the atom is initially prepared in a superposition of $|e\rangle$ and $|g\rangle$ states, we act with two resonant π Ramsey pulses such that the first pulse is tuned to the $g \leftrightarrow a$ transition frequency and the second pulse to $e \leftrightarrow g$. These two steps, in turn, transfer coherently the state of a qubit

$$\alpha|e\rangle + \beta|g\rangle \to \alpha|g\rangle + \beta|a\rangle, \quad \text{such that} \quad |\alpha|^2 + |\beta|^2 = 1 \quad (5.47)$$

CHAPTER 5: Generation of entangled states with a microwave cavity

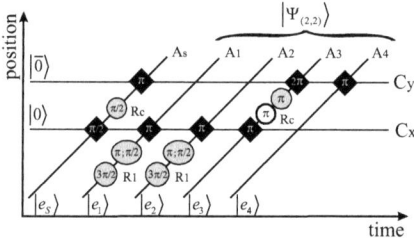

Figure 5.13: Temporal sequence that generates the 2×2 four-partite cluster state (5.41).

from $\{|e\rangle, |g\rangle\}$ into the $\{|g\rangle, |a\rangle\}$ basis. After this transfer is performed, the $A - C_y$ system undergoes a 2π Rabi pulse that yields the transformations (5.46), or equivalently, a controlled-z gate between the cavity mode C_y and an atomic qubit. Let us note, moreover, that in the experiments by S. Haroche and coworkers, the cavity has a small hole in the center of upper cavity mirror which enables one to couple a microwave source S' to an atom that moves through the cavity [see Fig. 5.1(a)]. This microwave source, therefore, can be used to act successively on both, the $g \leftrightarrow a$ and $e \leftrightarrow g$ atomic transitions and implement the coherent transfer (5.47).

The last operation we should discuss, is the controlled-z gate that acts upon the $C_x - C_y$ system and which is displayed in Fig. 5.11(d) in the terms of ancilla qubits. This gate acts on the cavity modes prepared in the state $|+, \bar{+}\rangle$ and produces the entangled state

$$\frac{1}{2}\left[|(0+1), \bar{0}\rangle + |(0-1), \bar{1}\rangle\right]. \tag{5.48}$$

In fact, the state (5.48) can be alternatively generated from the initially empty cavity $|0, \bar{0}\rangle$ by means of one auxiliary atom A_s initialized in the excited state and which crosses the cavity before the main chain of atoms. Specifically, the auxiliary atom first interacts for a $\pi/2$ Rabi pulse with the mode C_x, then for a $\pi/2$ Ramsey pulse with the microwave source S', and finally for a π Rabi pulse with the mode C_y. Using the Eqs. (5.12), (1.74) and (1.75), this sequence generates the state

$$\frac{1}{2}\left[|(0+1), \bar{0}\rangle + i\,|(0-1), \bar{1}\rangle\right], \tag{5.49}$$

while the auxiliary atom A_s is factored out in its ground state. In contrast to expression (5.48), the expression (5.49) contains one extra i factor that appears due to the orthogonal polarization of mode C_y with regard to C_x. This imaginary factor, however, later on is compensated in our scheme by the $A_{2N} - C_y$ mapping operation (π Rabi pulse) as we shall see below.

5.3. Two-dimensional cluster states

With this analysis of individual (gate) operations, we have determined all the ingredients which are needed to generate the $2 \times N$ cluster state, and which are entirely adapted to our cavity setup. The overall scheme is displayed in Fig. 5.12(b) in which the gates inside the dashed box must be repeated $N-3$ times. In addition to the notation we have used before, the letter R_c denotes the microwave field zone inside the cavity that is associated with the microwave source S' and the white circle denotes a Ramsey pulse that is tuned to the atomic $g \leftrightarrow a$ transition frequency. Note that the last $A_{2N}-C_y$ gate (π Rabi pulse) maps the cavity states $|\bar{0}\rangle$ and $|\bar{1}\rangle$ upon the atomic $|g_{2N}\rangle$ and $|e_{2N}\rangle$ states. According to the Eq. (1.75a), moreover, this mapping yields one extra i factor if the cavity mode C_y was empty, which together with the i factor from Eq. (5.49), produces an irrelevant global phase factor.

To understand better the scheme for the generation of $2 \times N$ cluster state, Fig. 5.13 displays the temporal sequence with all the steps which are needed to generate the 2×2 box state

$$|\Psi_{(2,2)}\rangle = \frac{1}{2}(|g_1,+_2,a_3,+_4\rangle + |g_1,-_2,g_3,-_4\rangle + |e_1,-_2,a_3,-_4\rangle + |e_1,+_2,g_3,+_4\rangle) \quad (5.50)$$

from Fig. 5.11(b). For the sake of brevity, the state $|g_s; 1, \bar{1}\rangle$ of auxiliary atoms and the cavity is not shown in this expression since they are both factored out after the sequence of steps is completed. Obviously, the state (5.50) is equivalent to the state (5.41) by making the assignments $\{|a_3\rangle = |\uparrow_3\rangle, |g_3\rangle = |\downarrow_3\rangle\}$ and $\{|g_i\rangle = |\uparrow_i\rangle, |e_i\rangle = |\downarrow_i\rangle\}$, where $i = 1, 2, 4$. In this section, therefore, we have shown that each atom in a chain of $2N$ (initially) uncorrelated atoms is incorporated into the $2 \times N$ cluster state by performing a Rabi π (followed by a 2π) pulse and, if required, also Ramsey pulses applied before and/or inside the cavity [61].

5.3.2 3×N and arbitrary two-dimensional cluster states

In section 5.1.3, we explained how to generate a $1 \times N$ cluster state by using one single cavity mode and in the previous section we explained how to generate a $2 \times N$ cluster state by using two cavity modes. Since the cavity supports only two modes corresponding to the different polarizations, one single cavity is not sufficient to generate a $3 \times N$ cluster state that would require three ancilla qubits.

In this section, instead, we shall present and explain a scheme that enables one to generate $3 \times N$ cluster state by using an array of two cavities, i.e. simply by placing one additional cavity $C^{(2)}$ together with Ramsey plates R_3 behind the plates R_2 and right before the detection area (the Ramsey plates R_d followed by detector) as displayed in Fig. 5.15(a). The scheme we like to present can be divided into the following two parts: (i) implementation of the controlled-z gates (edges) within a chain of $3N$ atoms according to Fig. 5.14(a) which leads to the generation of a $2 \times N$ displayed in Fig. 5.14(b). Notice

CHAPTER 5: Generation of entangled states with a microwave cavity

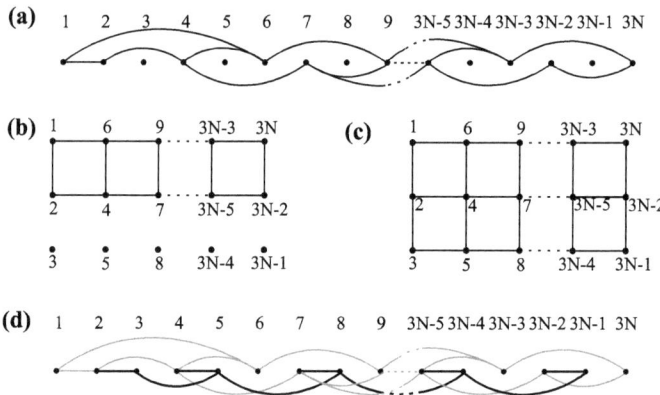

Figure 5.14: (a) Procedure for defining the edges for a chain of $3N$ atoms (nodes) such that an effective two-dimensional $2 \times N$ cluster state (b) is generated, where the nodes from the third row are disconnected from all other atoms of chain. (d) Definition of the remaining edges which transform the ($2 \times N$ cluster) state of atomic chain into a $3 \times N$ cluster state (c) by making use of the second cavity in setup.

that the third row of N atoms remains disconnected from all other $2N$ atoms and only the cavity $C^{(1)}$ and microwave sources S_1 and S_1' are utilized in this part. The circuit that generates these edges is the same as displayed in Fig. 5.12(c) up to a reassignment of the atomic labels. All the atoms that remain disconnected during this step simply pass through the first cavity being detuned from the resonance ($\Delta = \delta$) with both cavity modes. (ii) The second cavity $C^{(2)}$ and microwave sources S_2 and S_2' are then utilized in order to generate additional edges according to Fig. 5.14(d). This step completes the generation of a $3 \times N$ cluster state in which all neighboring nodes are connected to each other as displayed in Fig. 5.14(c).

Neither of these two steps do require any additional atom-cavity gates which have not been described and discussed previously. Therefore, the above procedure enables to generate $3 \times N$ cluster states via scheme that is well adapted for our setup with two cavities $C^{(1)}$ and $C^{(2)}$. The quantum circuit that accomplishes this task is displayed in Fig. 5.15(b), where the gates inside dash-boxed area must be repeated $N-4$ times. Similar as before, this circuit can be readily translated into a temporal sequence describing individual atom-cavity interactions and single atomic rotations. By having understood the construction of the $3 \times N$ cluster states, furthermore, the recipe which enables us to generate a two-dimensional (rectangular) cluster states of arbitrary size can be elaborated. Similarly as the $3 \times N$ cluster state has been generated from a $2 \times N$ cluster

5.3. Two-dimensional cluster states

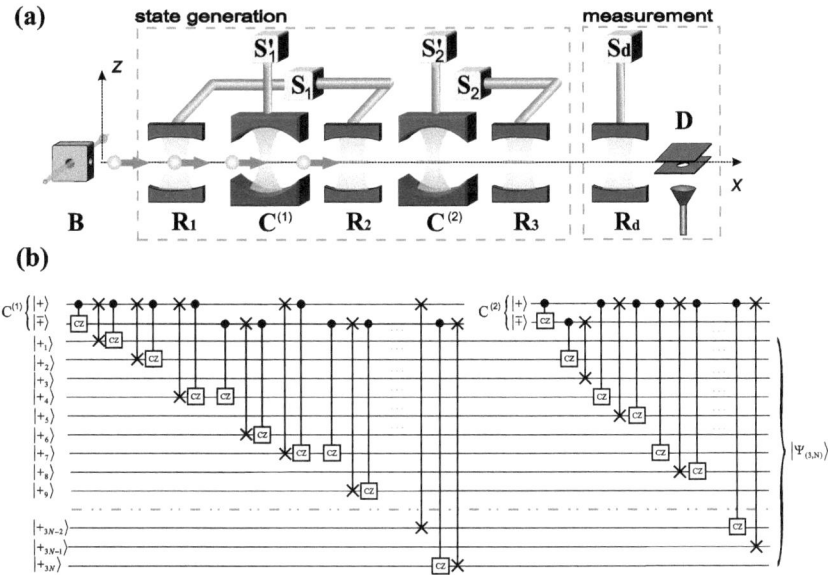

Figure 5.15: (a) Schematic setup of an experiment with two cavities C^1 and C^2. The classical field injected in Ramsey zones located before and after the cavities (as well as those inside of the cavities) are generated by the microwave sources S_1, S'_1, S_2 and S'_2. (b) Quantum circuit for generation of the $|\Psi_{(3,N)}\rangle$ cluster state between $3N$ initially uncorrelated qubits. In this circuit, two cavities $C^{(1)}$ and $C^{(2)}$ are utilized.

and N disconnected qubits, one can insert more cavities in the experimental setup and generate $M \times N$ cluster state (by means of $M - 1$ cavities in total), and where a proper assignment of atomic labels to the nodes of cluster state has to be made.

Of course, there may arise the question of how many atoms can be incorporated in a given cluster state. To obtain some rough estimation, let us consider a scenario in which each atomic qubit can be built into the entangled state for the price of a 3π Rabi pulse. If we assume, moreover, that the (minimum) distance between any two subsequent atoms in the chain is equal to the double waist length of cavity, then the number of atoms is related approximately to the coherent 'lifetime' T of atom-cavity system via expression

$$N \simeq \frac{1}{6}\frac{T}{T_\pi}\varepsilon, \qquad (5.51)$$

where T_π denotes the time required for realization of a single π Rabi pulse and ε is a factor which accounts for all corrections to our idealized scheme. Such corrections include the imperfect realization of Rabi and Ramsey pulses, the overlapping interaction of two

CHAPTER 5: Generation of entangled states with a microwave cavity

atoms from the chain with the same cavity mode, stray fields, and others. In practise, these effects leads to a much smaller number N of atoms which can be treated coherently. For the typical atomic velocities $v = 500\,m/s$ that is utilized in the experiments by S. Haroche and coworkers, a single π Rabi rotation takes about $T_\pi \approx 10\,\mu s$. The lifetime of atom-cavity system, in turn, is limited mainly by the radiative lifetime of circular Rydberg atoms $T \simeq 30\,ms$ (while the cavity relaxation time $\simeq 130\,ms$ is much longer [71]). Therefore, by making a conservative estimate of correction factor in Eq. (5.51), for instance $\varepsilon = 0.2$, we still obtain $N \simeq 100$ atoms which may pass through the setup.

5.4 Remarks on the implementation of schemes

Obviously, the generation of multi-partite entangled states with a trustworthy fidelity is an experimental challenge by itself. In the earlier cavity-QED experiments by A. Rauschenbeutel and coworkers [41], the generation of three-partite GHZ state (5.7) was reported with a fidelity of 0.54 %, that is just above the threshold which is necessary to prove the generation of this state. In this section, therefore, we shall discuss the main limitations which arise in cavity-QED experiments and which prevent the generation of large entangled states

One of the main bottlenecks in cavity-QED experiments is the quality of cavity mirrors, i.e. the presence of local roughness and deviations from the spherical shape. These defects cause the scattering of photons outside the cavity mode and thus reduce the coherent storage time of photons inside the cavity. The storage time of photons, in turn, limits the number of quantum operations (gates) that can be performed successively before the composite atom-cavity state becomes destroyed. This rapid loss of coherence due to the cavity relaxation, has stimulated S. Haroche and coworkers to develop a new ultrahigh-finesse cavity devices [71] for which the quality factor of mirrors has been increased by about two orders of magnitude. Such a high quality factor enables to perform more than hundred quantum logical operations within the lifetime of a photon inside the cavity (see estimations in the previous section).

In practice, the atoms which are emitted from the atomic source have a spread in their velocities. This spread leads to small deviations in the atom-cavity interactions times and it introduces uncertainties in the duration of Rabi and Ramsey pulses. In order to minimize this source of spread, a velocity selector has been placed right after the atomic source in setup by S. Haroche and co-workers. This selector reduces the velocity spread to ~ 2 m/s [35] which being compared to the typical velocity of 500 m/s of atoms, implies that the error due to the velocity spread is less than one percent and is negligible for the most purposes. This small spread in the velocities, in turn, gives rise also to a small spatial dispersion ($\lesssim 1$ mm) of atomic positions while passing through the cavity. This spatial dispersion compared to the resonant cavity wavelength (~ 5.9 mm), implies that only a small deflection from the cavity antinode may occur and yields as well a small deviation of atom-cavity coupling from its nominal value.

Indeed, the control and manipulation of cavity resonant frequency is essential in order to achieve the resonant atom-cavity interaction regime. Any deviation from this resonant regime would lead to spurious matrix elements, for instance, in the atom-cavity gates (5.13) and (5.42). In the experiments by S. Haroche and co-workers, the cavity frequency is manipulated by changing slightly the distance between two cavity mirrors. By making use of a piezoelectric stack placed under the lower cavity mirror, furthermore, a fine tuning of cavity length was achieved that corresponds to the frequency range of ~ 1 MHz, and which has to be compared to the atomic transition frequency 51.099 GHz. In this manuscript, however, we considered the scenario in which the atomic transition frequency is tuned to one of C_x or C_y cavity mode by applying a time-varying electric field across the cavity gap such that the required (Stark) shift of atomic $e \leftrightarrow g$ transition frequency is obtained. Instead of the instantaneous (step-like) change of atom-cavity detuning as displayed in the lower part of Fig. 5.1(b), a rather smooth switch of atom-cavity detuning is produced within the finite time of $\sim 1\,\mu s$ which could affect the evolution of cavity states, whenever the switching pulse is comparable to the applied Rabi pulse [77].

Finally, in order to describe a realistic evolution of atom-cavity system, one have to include also the interaction with the environment that has been omitted from the present considerations. In order to avoid the events with several atoms interacting with the same cavity mode, a sufficiently large spacing between the atoms passing through the cavity is necessary. This large spacing, in turn, implies that the cavity is empty during some time intervals and thus the field relaxation have to be taken into account. A detailed investigation of freely evolving cavity modes which interact with environment has been performed in Ref. [78] and the dynamics of cavity relaxation has been well understood.

5.5 Summary

In this chapter, various schemes have been suggested to (i) generate multi-partite GHZ and W states, (ii) generate one- and two-dimensional cluster states of arbitrary size, and (iii) prove the entanglement formation of three- and four-partite GHZ and W states. These schemes are based on the resonant interaction of a chain of Rydberg atoms with one or more cavities which support two (orthogonally polarized) modes of photon field. By using the graphical language of temporal sequences and quantum circuits, a comprehensive description of all necessary Rabi and Ramsey pulses together with all atomic manipulations has been achieved. Our goal is to provide the schemes which can readily

CHAPTER 5: Generation of entangled states with a microwave cavity

be adapted to the present-day microwave cavity-QED experiments although their realization is still a challenge, especially, if one is interested in entangled states with $N > 3$ atomic qubits involved.

Since the experimental reports [42, 69, 70], the use of cavities which support two independent cavity modes has been found an important step towards the generation and control of complex quantum states. A number of proposals [58, 72, 73, 74, 75, 76, 77] have been suggested in the literature to exploit further capabilities of such cavities and, in particular, engineering of various entangled states between the atomic (chain) and/or cavity qubits. In contrast to this work, however, most of the these suggestions are not well adapted to the recent developments in cavity-QED and no satisfactory attempt was made to reveal the non-classical correlations belonging to the produced entangled states.

The results of sections 5.2 and 5.3, moreover, suggest that cavity-QED provides a suitable framework not only for the generation of cluster states but also for one-way quantum computations which are performed by a sequence of single-qubit projective measurements (with possible feedforwarding). For this model of computations, a two-dimensional cluster state of appropriate size is required together with two types of measurements [52]: (i) measurement in the longitudinal basis $\{|\uparrow\rangle, |\downarrow\rangle\}$ and (ii) measurement in the transversal basis

$$\{(|\downarrow\rangle + e^{i\varphi}|\uparrow\rangle)/\sqrt{2}; \quad (|\downarrow\rangle - e^{i\varphi}|\uparrow\rangle)/\sqrt{2}\}, \tag{5.52}$$

where the remaining qubits which are not projected (measured) encode the output quantum state.

To show how the one-way quantum computation fits into our discussion, let us reconsider the setup displayed in Fig. 5.1(a). In this figure, a particular cluster state is generated within a chain of atoms right after it passes through the cavity and Ramsey plates R_1 and R_2. The generated cluster state (encoded in the atomic chain) then enters into the detection region, where each (Rydberg) atom is projected upon one of its levels e, g, or a and by which, therefore, the measurement in the basis (i) is performed. As we explained in section 5.2, the Ramsey pulse $\hat{R}(\pi/2, \varphi)$ followed by the detection of an atom in the basis (i) is equivalent to a projective measurement in the basis (5.52). We found in section 3.2.1, moreover, that the rotation $\hat{R}(\pi/2, \varphi)$ is efficiently realized by a sequence of three Ramsey pulses performed inside R_d, and where the first resonant pulse is followed by two short off-resonant pulses separated by a tunable time delay [see (3.10)]. The last (unmeasured) atoms in the atomic chain encode the final output quantum state. With this we can conclude that all necessary ingredients are available in order to perform one-way quantum computations in the framework of microwave cavity-QED and which includes also the preparation of required two-dimensional cluster state in the same setup.

Chapter 6

Generation of entangled states with an optical cavity

In the first part of this manuscript we introduced the off-resonant interaction regime between N circularly polarized three-level atoms and an optical cavity which supports two orthogonally polarized modes of photon field. We found that the cavity and a laser beam mediate together the interaction between atoms which are simultaneously coupled to them. We showed also that the evolution of initially uncorrelated atoms is described by the sequence (2.21) and is governed by the Hamiltonian (2.45). According to this Hamiltonian, moreover, the evolution of three-level atoms is reduced to the evolution of effectively two-level atoms which interact with each other via a two-photon exchange such that the fast decaying atomic states remain almost unpopulated. This energy exchange is quantitatively described by the W-class state (2.49) and is characterized by the complex amplitudes (2.52) which, in turn, are determined by atomic velocities and inter-atomic distances. By setting appropriate velocities of atoms and inter-atomic distances, therefore, one can generate the entangled W state (2.23) after the atomic chain leaves the cavity and decouples from both cavity and laser fields [62, 63].

In order to manipulate atomic velocities and inter-atomic distances, it is necessary (i) to place the atoms in a regular but adjustable linear lattice, (ii) to maintain a constant spacing between atoms for the entire time evolution, and (iii) to have an excellent control of motion of entire atomic chain. Nowadays, such a control is merely possible by using optical lattices (conveyor belts) introduced in section 4.2.1 and which have been recently utilized in various setups of cavity-QED [16, 18, 20]. The interference pattern produced by an optical lattice gives rise to a series of equidistant potential wells in which neutral atoms can be inserted and trapped. These wells, moreover, allow to control the inter-atomic separation and atomic velocities with a sub-micrometer precision over millimeter distances [80]. The experimental setup that combines optical cavity, a laser beam, and

CHAPTER 6: Generation of entangled states with an optical cavity

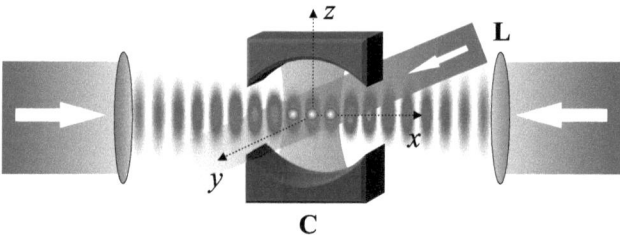

Figure 6.1: Schematic setup of an experiment in which a chain of trapped atoms is conveyed through cavity C and a laser beam L that acts perpendicularly to the cavity and lattice axes. See text for further discussions.

an optical lattice is displayed in Fig. 6.1 and it provides all the necessary ingredients to generate multipartite W entangled states between the atoms inserted in a lattice and which are conveyed through the cavity.

We finally mention that the off-resonant interaction regime is robust with regard to decoherence effects since the cavity mode and (fast decaying) excited atomic states remain almost unpopulated during the entire evolution. This robustness, in turn, plays one crucial role in the generation of multipartite entangled W states between atomic qubits encoded into the level structure of neutral atoms, and where the cavity plays the role of a data bus that mediates the interaction between these atomic qubits.

6.1 W states for atoms conveyed through a cavity

In section 2.2 we suggested a scheme to generate multipartite W-class states for initially uncorrelated atoms which are coupled simultaneously to the cavity and laser fields as displayed in Fig. 2.1(c). We found that by setting appropriate atomic velocities and inter-atomic distances, the generated W-class state reduces to the W state after the atomic chain leaves the cavity and decouples from both cavity and laser fields. As we explained in the beginning of this chapter, however, an excellent control of inter-atomic distances and the velocity of atomic chain is merely possible by using an optical lattice. The setup from Fig. 2.1(c), therefore, should be replaced by the setup displayed in Fig. 6.1, and this revised setup shall be considered as the starting point for our further discussions.

By this revised setup, a chain of N (initially uncorrelated) atoms is inserted into the sites of an optical lattice (being equally separated by a spacing d) and the entire chain is transported by the lattice with a constant velocity v along the x-axis such that atomic trajectories cross the cavity at the antinode. By this assumption, the position of each atom is described by the vector $\mathbf{r}_i(t) = \{x_i^o + v\,t, 0, 0\}$, where x_i^o denotes the initial position of i-th atom ($i = 1, \ldots, N$). As we briefly outlined in section 4.2.1, moreover, the velocity v of atomic chain is controlled by the shift in the frequencies of two counter-propagating laser beams while the inter-atomic distance d is manipulated by setting a proper wavelength of optical lattice.

Each of N identical (three-level Λ-type) atoms encodes a qubit in the metastable state $|0\rangle$ and the ground state $|1\rangle$ as displayed in Fig. 2.1(a). The chain of such atoms is prepared initially in the product state $|1_1, 0_2, \ldots, 0_N\rangle$ and it evolves according to the sequence (2.19) once it couples simultaneously to the cavity and laser fields. This sequence, in turn, reduces to the simplified sequence (2.21) if the atom-cavity and atom-laser couplings satisfy the conditions (2.18), (2.20), and (2.36). The evolution of composite atomic state $|1_1, 0_2, \ldots, 0_N\rangle$, therefore, is described by the wave-function (2.49) and is completely characterized by the complex amplitudes (2.52) which, in turn, are completely determined by atomic velocities and inter-atomic distances, once the frequency shifts: Δ_L, Δ_C, coupling constants: $g_\circ$, $\Omega_\circ$, and waists: w_L, w_C are fixed by a particular experimental setup. The wave-function (2.49), however, has not the desired form of a W state (2.23). In the next sections, we shall discuss the properties of W_N states for different values of N and calculate those v and d parameters, for which the function (2.49) becomes identical (or close) to the desired W states.

6.1.1 $N = 2$ partite state

For a chain of just two atoms ($N = 2$), the wave-function (2.49) takes the form

$$|\Phi_2(v,d)\rangle = e^{i\eta}\left(\cos\theta(v,d)\,|\mathbf{V}_1\rangle - i\,\sin\theta(v,d)\,|\mathbf{V}_2\rangle\right), \tag{6.1}$$

where $\eta = \sqrt{\frac{\pi}{2}}\,\frac{\Omega_\circ^2\, w_L}{4\Delta_L\, v}$ and $\theta(v,d)$ is given by the expression (2.51). From the wave-function (6.1), we readily recognize that (up to a global phase factor) the two-partite entangled states

$$|\Psi_2^\pm\rangle = \frac{1}{\sqrt{2}}\left(e^{\pm i\frac{\pi}{2}}|1_1,0_2\rangle + |0_1,1_2\rangle\right) \tag{6.2}$$

are obtained for (v,d) pairs fulfilling the condition $\theta(v,d) = (2n+1)\,\pi/4$ with n being an integer. Therefore, the maximal two-partite entangled state is obtained along the solid lines displayed in Fig. 6.2(a) for $n = 0, 1, 2, 3$, where velocities are displayed in the units of $q_\circ^2\, w_C/(\Delta_C\,\Delta_L^2)$ and distances in the units of w. Obviously, the change between maximally entangled (solid lines) and completely disentangled states (dashed lines) happens more and more rapidly as the velocity of atomic chain decreases from $n = 0$ onwards.

Apart from understanding the dynamical parameters (v,d) for which a maximum entanglement is achieved, it is important to know how sensitive these states are with

CHAPTER 6: Generation of entangled states with an optical cavity

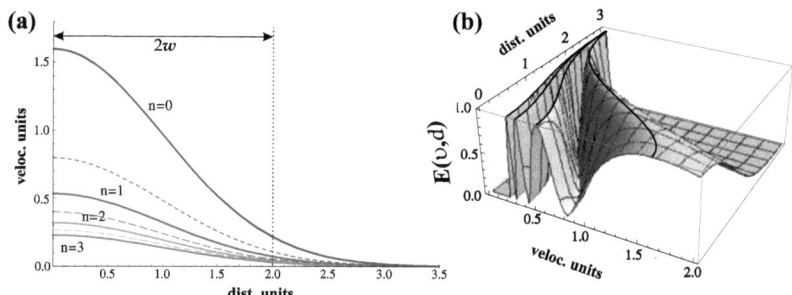

Figure 6.2: Atomic velocities v and inter-atomic distances d for which the condition $\theta(v,d) = (2n+1)\pi/4$ is satisfied, i.e., the initial product state of two atoms $|1_1, 0_2\rangle$ becomes maximally entangled (solid lines). The dashed lines, in contrast, indicate the (v,d) pairs for which the atomic qubits remain disentangled. (b) Plot of von Neumann entropy $E(v,d)$ as a function of atomic velocity and distance. In all these figures, the velocities v are displayed in units of $q_c^2 \, w_C/(\Delta_C \, \Delta_L^2)$ and the distances d in units of w.

regard to small uncertainties in the velocity and inter-atomic distance. To analyze this sensitivity, Fig. 6.2(b) displays the *von Neumann entropy* [4]

$$\begin{aligned}E(v,d) &= -\text{Tr}\left[\rho(v,d)\log_2 \rho(v,d)\right] \\ &= -\cos^2\theta(v,d)\log_2\left[\cos^2\theta(v,d)\right] - \sin^2\theta(v,d)\log_2\left[\sin^2\theta(v,d)\right] \end{aligned} \quad (6.3)$$

plotted for velocities and distances satisfying $\theta(v,d) < 2\pi$, and where $\rho(v,d)$ denotes the reduced density operator of $|\Phi_2(v,d)\rangle$ with regard to the second qubit. As expected, the maximal values of von Neumann entropy, i.e., $E(v,d) = 1$, are obtained along the lines which are displayed in Fig. 6.2(a). Moreover, the least rapid variation in the maxima occurs along the $n = 0$ line and for rather small inter-atomic distances. For small atomic velocities or some larger inter-atomic distance, in contrast, a good control of the entangled $|\Phi_2(v,d)\rangle$ states becomes more and more difficult.

Fig. 6.2(a) implies that an entanglement between the atoms occurs even for inter-atomic distances which are larger than $2w$, i.e. twice the cavity waist. Indeed, in a high finesse cavity the Gaussian profile (2.14) approximates quite well the intracavity field and, thus, it is possible to generate an entangled state even for the atomic separation $d > 2w$. In practice, however, the cavity relaxation and the spontaneous atomic decay introduce certain limitations on the distance between the atoms, beyond which it is not possible to generate entangled state (6.2). In order to estimate this limitation, we consider the condition [81]

$$N\,g_o^2\,\exp\left[-2\,x^2/w^2\right]/(\kappa\gamma) > 1, \qquad (6.4)$$

6.1. W states for atoms conveyed through a cavity

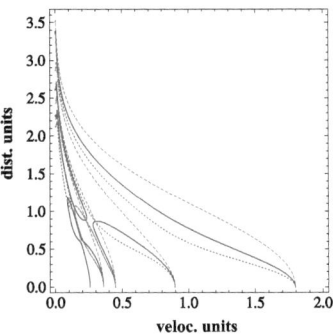

Figure 6.3: Lines along which the moduli $|C_1(v,d)|$ (solid), $|C_2(v,d)|$ (dashed), and $|C_3(v,d)|$ (dotted) are equal to $1/\sqrt{3}$. These lines correspond to velocities (6.11) with $n = 0, 1$ and $m = 1, 2$ (in the limit $d \to 0$). As before, velocities v are displayed in units of $q_o^2\, w_C/(\Delta_C\, \Delta_L^2)$ and the distances d in units of w.

which ensures that N atoms couple strongly to the cavity field and, therefore, implies the validity of effective evolution (2.45). Here, κ and γ denote the cavity loss rate and the atomic decay rate, respectively. For $N = 2$, furthermore, the above condition bounds the atomic coordinate to the interval $x_- < x < x_+$ with

$$x_\pm = \pm w \sqrt{\frac{\ln[2\, g_o^2/\,(\kappa\,\gamma)]}{2}}. \tag{6.5}$$

Owning to these boundaries, therefore, the distance d between two atoms should not exceed

$$\frac{d}{w} < \sqrt{2\,\ln[2\, g_o^2/\,(\kappa\,\gamma)]} = \frac{x_+}{w} - \frac{x_-}{w}. \tag{6.6}$$

For the typical atom-cavity parameters [79]: $\{g_o, \kappa, \gamma\} = 2\pi \times \{10, 0.4, 2.6\}$ MHz, this condition implies the limitation $d < 3.243\, w$. We note that this estimation agrees well with the solid lines from Fig. 6.2(a) since, for $d > 3.2\, w$, the atomic velocity becomes negligibly small and prevents any experimental implementation of our scheme.

6.1.2 $N = 3$ partite state

For a chain of three atoms ($N = 3$), the wave-function (2.49) takes the form

$$|\Phi_3(v,d)\rangle = e^{i(\eta-\zeta)}\left(C_1(v,d)|\mathbf{V}_1\rangle + C_2(v,d)|\mathbf{V}_2\rangle + C_3(v,d)|\mathbf{V}_3\rangle\right), \tag{6.7}$$

99

CHAPTER 6: Generation of entangled states with an optical cavity

with

$$C_1(v,d) = \frac{-\xi^3 \lambda_- + \sqrt{8+\xi^6}\,(\lambda_+ + 2e^{i\varsigma})}{4\sqrt{8+\xi^6}}, \qquad C_3(v,d) = \frac{-\xi^3 \lambda_- + \sqrt{8+\xi^6}\,(\lambda_+ - 2e^{i\varsigma})}{4\sqrt{8+\xi^6}},$$

$$C_2(v,d) = -\frac{\lambda_-}{\sqrt{8+\xi^6}}, \qquad (6.8)$$

and where we used the notation

$$\xi = \exp\left[-\frac{d^2}{2w_C^2}\right], \quad \lambda_\pm = \exp\left[i\,2\,\xi\chi\sqrt{8+\xi^6}\right] \pm 1, \quad \chi = \sqrt{\frac{\pi}{8}\frac{q_0^2\,w_C}{\Delta_C\,\Delta_L^2}\,v},$$

$$\varsigma = \xi\chi\left(3\xi^3 + \sqrt{8+\xi^6}\right), \quad \zeta = \xi\chi\left(\xi^3 + \sqrt{8+\xi^6}\right).$$

In order to obtain the W_3 state from the wave-function $|\Phi_3(v,d)\rangle$, we have to determine those (v,d) pairs for which the equations

$$|C_1(v,d)| = |C_2(v,d)| = |C_3(v,d)| = \frac{1}{\sqrt{3}} \qquad (6.9)$$

are fulfilled. In Fig. 6.3(a), we displayed the corresponding lines for which the moduli $|C_1(v,d)|$ (solid), $|C_2(v,d)|$ (dashed), and $|C_3(v,d)|$ (dotted) are equal to $1/\sqrt{3}$. The W_3 state is obtained for those (v,d) pairs, therefore, for which all three types of lines intersect with each other.

As seen from Fig. 6.3(a), however, the lines for the (moduli of) amplitudes $C_i(v,d)$ intersect only in the case of vanishing inter-atomic distance. In order to determine the corresponding velocities, we first observe that for $d \to 0$ ($\xi \to 1$), the wave-function (6.7) becomes

$$e^{i(\eta-4\chi)}\frac{1+2e^{i\,6\chi}}{3}|\mathbf{V}_1\rangle + e^{i(\eta-4\chi)}\frac{1-e^{i\,6\chi}}{3}(|\mathbf{V}_2\rangle + |\mathbf{V}_3\rangle),$$

which (up to a global phase factor) can be readily cast into the W_3 form

$$|\Psi_3^\pm\rangle = \frac{1}{\sqrt{3}}\left(e^{\pm i\frac{2\pi}{3}}|\mathbf{V}_1\rangle + |\mathbf{V}_2\rangle + |\mathbf{V}_3\rangle\right) \qquad (6.10)$$

if $\chi = (3n+m)\pi/9$ or, equivalently, if the velocity takes the values

$$v = \sqrt{\frac{\pi}{8}\frac{q_0^2\,w_C}{\Delta_C\,\Delta_L^2}}\,\frac{9}{\pi(3n+m)}, \qquad (6.11)$$

with $m = 1,2$ and n being integers. To summarize, the vanishing inter-atomic distance along with velocities (6.11) are necessary to obtain the W_3 states due to wave-function (6.7).

According to our setup from Fig. 6.1, however, the atoms are separated by a non-zero distance which is non-negligible with regard to the cavity waist. Therefore, we shall determine the (v,d) pairs with non-vanishing distance, for which the probability to

6.1. W states for atoms conveyed through a cavity

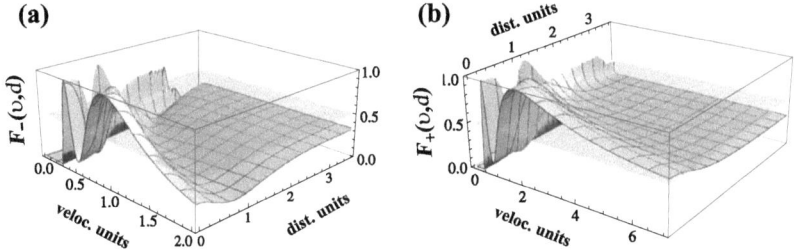

Figure 6.4: (a) Two maxima in the fidelity $F_-(v,d)$ obtained for velocities (6.11) with $n = 0, 1$ and $m = 1, 2$ (in the limit $d \to 0$). For guidance of eyes, the semi-transparent layer displays a constant value $F_-(v,d) = 0.5$. (b) The same as in figure (a) but for the fidelity $F_+(v,d)$. Again, velocities v are displayed in units of $q_o^2 \, w_C/(\Delta_C \, \Delta_L^2)$ and the distances d in units of w.

obtain the state (6.10) due to the wave-function (6.7) is highest possible. To determine such (v,d) parameter region, we utilize the fidelities [4]

$$F_\pm(v,d) = |\langle \Psi_3^\pm | \Phi_3(v,d) \rangle|^2 = \frac{1}{3} \left| e^{\mp i \frac{2\pi}{3}} C_1(v,d) + C_2(v,d) + C_3(v,d) \right|^2, \quad (6.12)$$

which are the measures of distance between the states $\{|\Psi_3^+\rangle, |\Phi_3(v,d)\rangle\}$ and between the states $\{|\Psi_3^-\rangle, |\Phi_3(v,d)\rangle\}$. These fidelities are displayed in Figs. (6.4)(a),(b) together with semitransparent planes in order to delimit the regions for which $F_\pm(v,d) \geq 0.5$. While the maximum values $F_\pm(v,d) = 1$ are obtained only for vanishing inter-atomic distances, there are still v and d pairs (with non-zero distance) for which the fidelities become reasonably close to the maximal value. Note that the region for which $F_+(v,d) \geq 0.5$ is notably larger than those for which $F_-(v,d) \geq 0.5$. We conclude, therefore, that from the experimental perspective it might be preferable to generate the state $|\Psi_3^+\rangle$ since the respective (v,d) region is notable larger as compared to the state $|\Psi_3^-\rangle$.

6.1.3 $N = 4$ partite state

For a chain of four atoms ($N = 4$), the wave-function (2.49) can be written as

$$|\Phi_4(v,d)\rangle = e^{i(\eta - \zeta)} \left(C_1(v,d)|\mathbf{V}_1\rangle + C_2(v,d)|\mathbf{V}_2\rangle + C_3(v,d)|\mathbf{V}_3\rangle + C_4(v,d)|\mathbf{V}_4\rangle \right). \quad (6.13)$$

In contrast to $N = 2$ or $N = 3$ cases, however, the expressions for amplitudes $C_i(v,d)$ become rather bulky now and are not displayed here. Recall from the previous section that the wave-function $|\Phi_3(v,d)\rangle$ reduced to the $|\Psi_3^\pm\rangle$ states only for vanishing distances and velocities (6.11). In this section, therefore, we proceed in a similar fashion and

CHAPTER 6: Generation of entangled states with an optical cavity

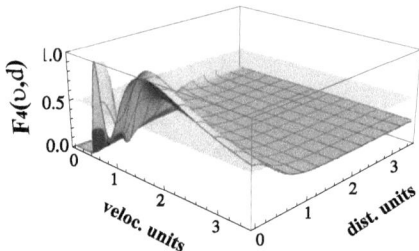

Figure 6.5: Fidelity $F_4(v,d)$ for the generation of W_4 state as a function of velocity and the inter-atomic distance. Again, the maximum value $F_4(v,d) = 1$ is obtained only for vanishing distance ($d = 0$) and velocities (6.15) with $n = 0, 1$. The semi-transparent plane with $F_4(v,d) = 0.5$ is plotted to guide the eyes and the units are the same as in previous figures.

consider first the wave-function (6.13) in the limit of vanishing inter-atomic distance ($d \to 0$)

$$e^{i(\eta-4\chi)}\frac{1+3e^{i\,8\chi}}{4}|V_1\rangle + e^{i(\eta-4\chi)}\frac{1-e^{i\,8\chi}}{4}\sum_{j=2}^{4}|V_j\rangle.$$

From this expression, (up to as global phase factor) the W_4 state

$$|\Psi_4^W\rangle = \frac{1}{2}\left(e^{i\,\pi}|V_1\rangle + |V_2\rangle + |V_3\rangle + |V_4\rangle\right) \quad (6.14)$$

is readily produced if $\chi = \pi(2n+1)/8$ or, equivalently, if the velocity takes the values

$$v = \sqrt{\frac{\pi}{8}\frac{q_o^2\,w_C}{\Delta_C\,\Delta_L^2}\frac{8}{\pi(2n+1)}}. \quad (6.15)$$

Similarly to the previous section, in Fig. (6.5) we displayed the fidelity $F_4(v,d) = |\langle\Psi_4^W|\Phi_4(v,d)\rangle|^2$ for the generation of W_4 states as a function of velocity and inter-atomic distance due to wave-function (6.13). The maximum fidelity $F_4(v,d) = 1$ is obtained only for vanishing distance ($d = 0$) and velocities which fulfill the condition (6.15). For non-vanishing distances, nevertheless, there is one broad parameters region for which the W_4 state (6.14) can be generated with a reasonable hight fidelity.

6.1.4 $N \geq 5$ partite states

For any other number $N \geq 5$ of atoms in the chain, the amplitudes $C_1(v,d), \ldots, C_N(v,d)$ can be computed with help of expression (2.52), and the obtained wave-function

$$|\Phi_N(v,d)\rangle = e^{i(\eta-\zeta)}\sum_{i}^{N} C_i(v,d)|V_i\rangle \quad (6.16)$$

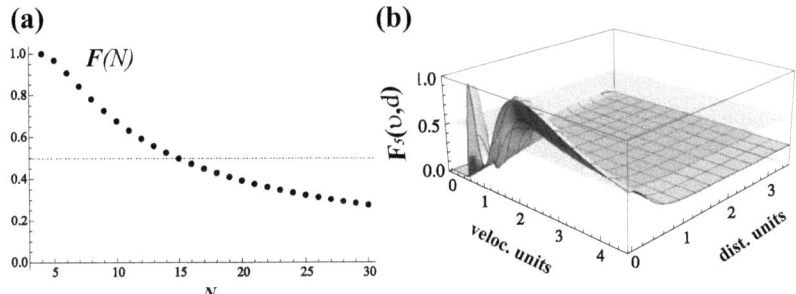

Figure 6.6: (a) The fidelity $F(N) = |\langle \Psi_N^W(\pi)|\Phi_N\rangle|^2$ has its maximum value $F(N) = 1$ for $N = 4$ and decreases monotonically as the number of atoms increases. For $N > 15$, it falls below the threshold $F(N) = 1/2$ (dotted line). (b) Fidelity $F_5(v,d)$ for the production of $|\Psi_5^W\rangle$ state due to $|\Phi_5(v,d)\rangle$ as a function of velocity and the interatomic distance. Similar as in previous figures, it reaches its maxima for $d = 0$ and velocities (6.21) with $n = 0, 1$.

can be further analyzed with regard to (v,d) pairs for which the corresponding W_N state is produced most reliably, i.e., with a highest possible fidelity. In order to proceed, however, we still need to specify the reference W_N state which we are looking in the form

$$|\Psi_N^W(\phi)\rangle = \frac{1}{\sqrt{N}}\left(e^{i\phi}|\mathbf{V}_1\rangle + \sum_{i=2}^{N}|\mathbf{V}_i\rangle\right), \qquad (6.17)$$

where ϕ is an unknown phase. The form of this state was chosen in line with the previously obtained W states (6.2), (6.10), and (6.14). In order to calculate the unknown phase ϕ, we follow the procedure form previous sections and consider the wave-function (6.16) in the limit $d \to 0$ ($\xi \to 1$)

$$|\Phi_N(v,0)\rangle = e^{i(\eta - 4\chi)} \sum_{k=1}^{N} \frac{1 + (\delta_{k1}N - 1)\exp(i\,2N\chi)}{N}|\mathbf{V}_k\rangle. \qquad (6.18)$$

The obtained wave-function depends only on the atomic velocity v that is encoded in the scalar function χ. The right hand part of wave-function (6.18) implies, moreover, that the amplitudes $C_k(v,0)$ cannot fulfill the (W_N) equalities

$$|C_1(v,0)| = \ldots = |C_N(v,0)| = \frac{1}{\sqrt{N}} \qquad (6.19)$$

for any value of v. However, we can find those values of χ for which the expressions

$$\left||C_1(v,0)| - \frac{1}{\sqrt{N}}\right|, \ldots, \left||C_N(v,0)| - \frac{1}{\sqrt{N}}\right| \qquad (6.20)$$

CHAPTER 6: Generation of entangled states with an optical cavity

become minimal. In other words, we determine those atomic velocities for which the wave-function (6.18) approximates the W_N state [given by the conditions (6.19)] as close as possible. It can be straightforwardly shown that all the expressions (6.20) are minimized for the values $\chi = \pi(2n+1)/(2N)$ or, equivalently, for velocities

$$v = \sqrt{\frac{\pi}{8} \frac{q_o^2 \, w_C}{\Delta_C \, \Delta_L^2} \frac{2N}{\pi(2n+1)}}. \tag{6.21}$$

By substituting the obtained value for χ in the wave-function (6.18), we obtain (up to a global constant phase) the parameter independent state

$$|\Phi_N\rangle = \frac{2-N}{N}|\mathbf{V}_1\rangle + \frac{2}{N}\sum_{i=2}^{N}|\mathbf{V}_i\rangle. \tag{6.22}$$

As we just mentioned, there are no such velocities for which the equalities (6.19) can be fulfilled. However, we found the state (6.22) which gives the best approximation to the W_N state and, at the same time, it is obtained from the wave-function (6.16) by means of velocities (6.21). On the other hand, we have specified the explicit form of reference W_N state (6.17) except the unknown phase ϕ. By comparing the state $|\Psi_N^W(\phi)\rangle$ for $N=4$ with the state $|\Phi_N\rangle$, we find that the phase ϕ is equal to π. In order to understand how well the state (6.22) approximates the state $|\Psi_N^W(\pi)\rangle$, in Fig. 6.6(a) we displayed the fidelity $F(N) = |\langle \Psi_N^W(\pi)|\Phi_N\rangle|^2$. As seen from this figure, the fidelity has its maximum value $F(N) = 1$ for $N = 4$ and decreases monotonically as the number N of atoms is increases. The fidelity drops below the threshold $F(N) = 1/2$ for $N > 15$. We conclude, therefore, that the state (6.22) approximates reasonably well the reference state W_N only for $4 < N < 15$.

Having determined the reference W_N state, we can evaluate each fidelity

$$F_N(v,d) = |\langle \Psi_N^W(\pi)|\Phi_N(v,d)\rangle|^2; \quad 5 \leq N < 15 \tag{6.23}$$

as a function of velocity and inter-atomic distance. In Fig. 6.6(b), for instance, we display this fidelity for $N = 5$. According to this figure, moreover, the fidelity reaches its maxima $F_5(v,d) = F(5) \approx 0.97$ for $d = 0$ and velocities which satisfy the condition (6.21) with $n = 0, 1$. Let us note here that the typical spacing between two neighbored potentials wells (sites) of an optical lattice lies in the sub-micrometer range [16, 18, 20]. As seen from Fig. 6.6(b), this typical spacing is comparable to the inter-atomic distance for which the fidelity $F_5(v,d) \approx 0.9$ is reasonable high and where the typical cavity waist ($w = 20$ μm) has been considered as the distance units. The recent developments in optical cavity-QED, therefore, make it possible to generate W_5 state by means of proposed scheme. If we compare, furthermore, the (v,d) regions for which fidelities $F_\pm(v,d)$ [see Fig. 6.4], $F_4(v,d)$ [see Fig. 6.5], and $F_5(v,d)$ [see Fig. 6.6(b)] are higher than the threshold value of $1/2$, we can conclude that these regions become smaller as the number of atoms (in the chain) increases.

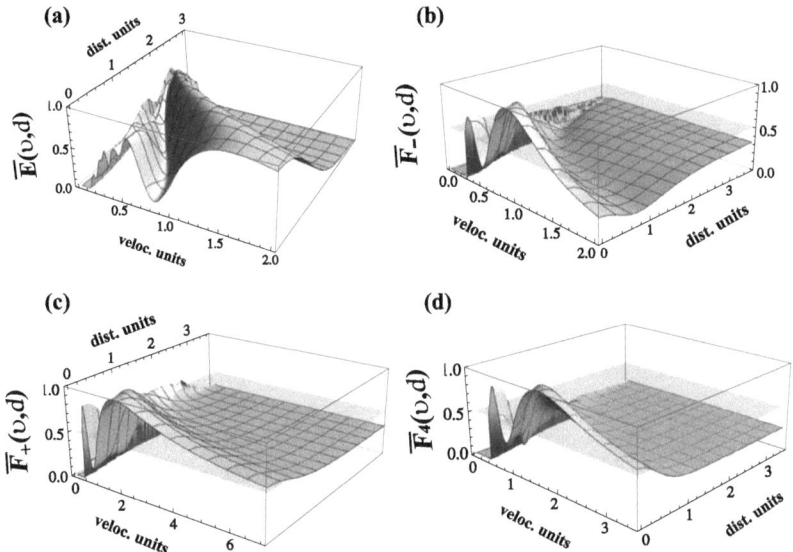

Figure 6.7: Entanglement and fidelity measures averaged over 20 randomly chosen uncertainties Δd and Δv of inter-atomic distance and the velocity, respectively. (a) Von Neuman entropy $\overline{E}(v,d)$, (b) fidelities $\overline{F}_-(v,d)$ and (c) $\overline{F}_+(v,d)$ for a chain of three atoms, and (d) $\overline{F}_4(v,d)$ for a chain of four atoms. The uncertainties for the distance and velocities are chosen from the intervals $[0,\,0.2\,\overline{d}]$ and $[0,\,0.1\,\overline{v}]$, respectively. See the text for further details.

6.2 Remarks on the implementation of schemes

In our discussions so far, we have always assumed that the velocity and the distance of atoms in chain, i.e. their position within the optical lattice, can be controlled exactly. With this assumption in mind, the atom-cavity and atom-laser couplings were described by the expression (2.24) and (2.25), respectively. This assumption, however, neglects the oscillations of atoms within the potential wells due to their finite temperatures which include both the axial (along the x-axis) and radial (along the y, z-axes) oscillations [see Fig. 2.1(c)]. This additional motion gives rise to a dispersion of atomic positions and velocities and leads to uncertainties of dynamical parameters in our model.

Obviously, any significant uncertainty in the parameters $\{v, d, \Delta_C, \Delta_L, g_\circ, \Omega_\circ\}$ will influence the generation of desired W entangled states. According to our scheme, however, these entangled states are generated when all the atoms have left the cavity. Instead of understanding these parameters as *exact*, therefore, they should refer to the mean values

CHAPTER 6: Generation of entangled states with an optical cavity

and we have to analyze how small (but realistic) variations in these parameters can affect the final state of atoms within the chain. For instance, the radial oscillations of atoms lead to the mean value of vacuum Rabi frequency $\overline{g}_\circ$ and axial oscillations to the mean values of inter-atomic distance $\overline{d}$ and velocity $\overline{v}$, respectively. Axial oscillations affects also the initial position x_i^o and velocity v_i of each atom inside the lattice and result in uncertainties $\triangle d_i = \overline{d} - |x_{i+1}^o - x_i^o|$ and $\triangle v_i = \overline{v} - v_i$, where $i = 1, \ldots, N$. Therefore, the plots $E(v,d)$, $F_\pm(v,d)$, $F_4(v,d)$, and $F_5(v,d)$ from Figs. 6.2(b), 6.4, 6.5, and 6.6(b) should be re-plotted as function of mean values $\overline{v}$ and $\overline{d}$ and their corresponding uncertainties, respectively.

In order to determine realistic uncertainties for the distance and velocity of atoms in chain, we first mention that the recent developments allow to position the atoms with respect to the cavity antinode with a precision of ~ 0.1 μm by utilizing an additional dipole trap acting along the cavity z−axis [16]. When compared to the typical cavity wavelength (~ 0.85 μm), such positioning precision leads to a spatial dispersion which, in turn, yields the mean value $\overline{g}_\circ \approx 0.7\, g_\circ$ and which is still good enough for our purposes [see inequalities (2.18) and (2.36)]. Moreover, the same spatial dispersion implies upper bounds for the uncertainties $|\triangle d/\overline{d}| \lesssim 0.2$ and $|\triangle v/\overline{v}| \lesssim 0.1$, if compared with the typical spacing (~ 0.5 μm) between two neighbored potential wells of an optical lattice and the typical atomic velocities (~ 0.5 m/s) along the lattice axis.

For a further analysis of how reliably a given (experimental) setup will generate a particular W state, in Fig. 6.7 we display the (mean) functions $\overline{E}(v,d)$, $\overline{F}_\pm(v,d)$, and $\overline{F}_4(v,d)$ by calculating their average for a certain spread of parameters. For each subfigure 6.7(a)-(d), we have randomly chosen 20 uncertainties $\triangle d$ and $\triangle v$ from the intervals $[0, 0.2\,\overline{d}]$ and $[0, 0.1\,\overline{v}]$, respectively. By comparing the Figs. 6.2(b) and 6.7(a) it can be seen that the von Neumann entropy, for instance, is reduced considerably for its sharp maxima ($n = 3$) and that it remains almost the same around the broad maxima ($n = 0$). Similarly, the mean fidelities which are displayed in Figs. 6.7(b)-(d), are considerably reduced for their sharp maxima. These (v, d) regions for the velocity and inter-atomic distance in the atomic chain are, therefore, less useful for any practical implementation and only the (v, d) regions which correspond to the broad maxima of von Neumann entropy and fidelities, are favorable for the generation of entangled W states by means of proposed scheme.

6.3 Summary

In this chapter, a scheme was proposed to generate entangled W states for a chain of N three-level atoms which are equally separated (inside the chain) and conveyed through an optical cavity by means of an optical lattice. This scheme is based on the cavity-

6.3. Summary

laser mediated interaction between the atoms which are separated by a macroscopic distance and it works in a completely deterministic way. Only two parameters, namely the velocity of chain and the inter-atomic distance determine the effective interaction of atoms and, therefore, the degree of entanglement that is obtained for the entire chain [62, 63]. The purpose of this chapter is to understand the state evolution of atomic chain and how it can be utilized to generate entangled W states. For chains that consists of $N = 2, 3, 4$ and 5 atoms, Figs. (6.2)-(6.6) display the von Neumann entropy and fidelities as functions of velocity and inter-atomic distance. For $5 \leq N < 15$, moreover, we suggested the reference state $|\Psi_N^W(\pi)\rangle$ which is approximated by the wave-function $|\Phi_N\rangle$ with a high fidelity. In view of the recent developments in optical cavity-QED, moreover, we have also analyzed and discussed the proposed scheme regarding the sensitivity in formation of desired entanglement due to uncertainties in the atomic motion.

For two or more atoms, the generation of entanglement by means of a (detuned) optical cavity has been investigated in several papers [21, 32, 86]. All these studies, however, relied on the *small sample* approximation in which the separation of the atoms is considered to be negligible if compared with the cavity waist. Only recently [87, 88, 89], the atom-cavity coupling (2.14) has been exploited in more detail in order to suggest various entanglement schemes within cavity-QED. In the work by M. Amniat-Talab, for instance, a scheme was proposed in which two atoms were coupled sequentially to a resonant cavity and where a position-dependent coupling is used to drive a STIRAP-type process in order to reduce the losses due to atomic and cavity decays. Moreover, the scheme by C. Marr is also based on a STIRAP-type process and describes an adiabatic evolution of a product state of two atoms which are coupled simultaneously to a detuned cavity. In both schemes, however, the atomic velocity and inter-atomic separation are used to control the accuracy of a STIRAP-type process, in contrast to our approach, in which these parameters are utilized to control the degree of entanglement.

Our proposed scheme might be suitable also for ion-cavity experiments in which N trapped ions interact simultaneously with a (detuned) optical cavity [90, 91]. In these experiments, the same coupling to the laser and cavity fields applies for ions with a three-level Λ-type configuration as displayed in Fig. 2.1(a). For such a level configuration, the W state can be generated by moving the equally distanced trapped ions (along the trap) through the cavity. Similarly as for the atomic chain above, the cavity-laser mediated interaction between the ions is described by the effective Hamiltonian (2.45) and, therefore, requires the same analysis as performed in this chapter in order to produce the entangled W_N states.

CHAPTER 6: Generation of entangled states with an optical cavity

Outlook and Acknowledgements

In this manuscript, we presented several practical schemes for generation of multipartite entangled states for a chain of atoms which pass through one or more high-finesse resonators. In the first step, we proposed two schemes for generation of one- and two-dimensional cluster states of arbitrary size. These schemes are based on the resonant interaction of a chain of Rydberg atoms with one or more microwave cavities. In the second step, we proposed a scheme for generation of multipartite W states. This scheme is based on the off-resonant interaction of a chain of three-level atoms with an optical cavity and a laser beam. We described in details all the individual steps which are required to realize the proposed schemes and, moreover, we discussed several techniques to reveal the non-classical correlations associated with generated small-sized entangled states.

In the first chapter, we introduced the interaction of a single atom with a single-mode monochromatic light field in the quantum regime. We derived the Hamiltonian that governs this interaction by assuming that the cavity supports two linearly and orthogonally polarized modes of light, while the atom emits or absorbs the circularly polarized light during its transition. We analyzed, furthermore, the situation in which the atomic transition frequency matches exactly the frequency of one of cavity modes and we found that the atom-cavity evolution is governed by the Jaynes-Cummings Hamiltonian. We derived the evolution for this Hamiltonian and we realized that this evolution describes the time-varying entanglement of a two-level atom with cavity photon field. We found, moreover, how the obtained atom-cavity evolution is affected by the uniform motion of an atom that probes one transversal component of cavity field amplitude. Finally, we analyzed the effects of spontaneous atomic emission and cavity relaxation in order to understand how the energy exchange of coupled atom-cavity system evolves in realistic environments.

In the second chapter, we introduced the interaction of a single atom with single-mode monochromatic light field in the semiclassical regime. We derived the Hamiltonian that governs this interaction and we found the respective evolution. We realized that the obtained evolution describes the oscillations of electronic population between its ground and excited states. Next, we explained our scheme for generation of atomic multipartite

OUTLOOK AND ACKNOWLEDGEMENTS

entangled states that is based on the off-resonant interaction regime of three-level atoms placed in the cavity and coupled simultaneously to a laser beam. By performing the adiabatic elimination procedure, we showed that the evolution of three-level atoms is reduced to the evolution of effectively two-level atoms which interact with each other via a two-photon exchange such that the fast decaying atomic excited states remain almost unpopulated. This energy exchange is quantitatively described by the W-class state and is determined by the atomic velocities and inter-atomic distances. By setting appropriate velocities of atoms and inter-atomic distances, therefore, we showed how to generate the entangled W state from the W-class state, after the atomic chain leaves the cavity and decouples from both cavity and laser fields.

In the third chapter, we defined the strong coupling regime of atom-cavity interaction and described in details the characteristics of highly excited Rydberg atoms and a microwave cavity – the basic constituents of setup developed in the group of S. Haroche. We also explained the role of Ramsey plates in manipulation of an arbitrary superposition of atomic states in question. In addition, we described the procedure that enables to set the phase of this superposition and which is based on the sequence of short off-resonant pulses. In the fourth chapter, furthermore, we described in details the characteristics of atoms with low-lying electronic states and an optical cavity – the basic constituents of setup developed in the group of D. Meschede. We also explained the working principle and role of an optical lattice (conveyor belt) utilized to transport the atoms through the cavity. We concluded, that the mentioned setups meet the strong coupling conditions and, therefore, provide all the necessary ingredients to generate complex entangled states of atomic qubits, where the cavity plays the role of a data bus.

In the fifth chapter, we first presented two schemes to generate multi-partite GHZ and W states. In addition, we described two techniques to reveal the non-classical correlations associated with two-partite entangled states. The first technique is based on the transversal measurements of two atoms, while the second is based on the free evolution of cavity modes. Using these techniques, we suggested the schemes which reveal the non-classical correlations of small-sized GHZ and W states. Next, we presented a scheme to generate the linear cluster state, and right afterwards, two schemes to generate the two-dimensional $2 \times N$ and $3 \times N$ cluster states. We showed how the last scheme can be extended to generate two-dimensional cluster states of arbitrary size, once a sufficiently large chain of atoms and an array of cavities are provided. For all these schemes, we described the individual steps in the interaction of each atom with one of cavity modes and we made use of graphical language in order to display all these steps in terms of quantum circuits and temporal sequences. We briefly discussed the implementation of our schemes by considering the setup similar to those utilized in the group of S. Haroche, and we concluded that cavity-QED provides a suitable framework not only for the generation of cluster states but also for one-way quantum computations.

In the last chapter, we proposed a scheme to generate entangled W states for a chain of N three-level atoms which are equally separated (inside the chain) and conveyed through an optical cavity by means of an optical lattice. This scheme is based on the cavity-laser mediated interaction between the atoms which are separated by a macroscopic distance and it works in a completely deterministic way. Only two parameters, namely the velocity of chain and the inter-atomic distance determine the effective interaction of atoms and, therefore, the degree of entanglement that is obtained for the entire chain. For chains that consists of $N = 2, 3, 4$ and 5 atoms, we discussed the properties of obtained states and calculated those velocities and distances, for which this state reduces to a given W_N states most reliably. Moreover, we displayed the von Neumann entropy and respective fidelities as functions of velocity and inter-atomic distance and we found preferable velocities and distances which can be useful for a practical implementation. Apart from generation of W states, we analyzed how robust are the generated entangled states with respect to small oscillations in the atomic motion. Finally, we discussed the implementation of our scheme by considering the setup similar to those utilized in the group of D. Meschede, and we concluded that our schemes can be adapted to the near-future developments in cavity QED.

Finally, I would like to take the opportunity and say a few words about the people who contributed to this work and helped me in one or the other form. First and foremost I wish to thank my supervisor Priv.-Doz. Dr. Stephan Fritzsche for giving me the opportunity to do my PhD in his group, and maybe even more, for a constant support that made my stay in Kassel and Heidelberg a quite worthwhile experience. It goes without saying that I have dramatically profited from his experience and intuition. I must say, moreover, that I appreciate the freedom I had to follow my own ideas and interests. Secondly, it is a pleasure for me to thank Dr. Thomas Radtke with whom it was always interesting and funny to collaborate. I would also like to thank Priv.-Doz. Dr. Jörg Evers, Prof. Dr. Markus Oberthaler, Prof. Dr. Thomas Gasenzer, and Prof. Dr. Peter Schmelcher for helpful discussions and their interest in my research project.

Apart from these persons, I wish also to thank Dr. Andrey Surzhykov and his excellent theory group. Working in the same environment, I participated in many interesting and inspiring discussions about physics and far beyond which contributed significantly to the nice working atmosphere and made my stay in Heidelberg a quite enjoyable time. Moreover, I wish especially thank my family and my friends for their encouragement that played an essential role for me during these years.

OUTLOOK AND ACKNOWLEDGEMENTS

Bibliography

[1] D. Deutsch, "Quantum theory, the Church-Turing principle and the universal quantum computer", Proc. Roy. Soc. Lond. A **400** 97 (1985).

[2] P. W. Shor, "Algorithms for quantum computation: Discrete logarithms and factoring", in Proceedings of the 35th Annual Symposium on the Foundations of Computer Science, (IEEE Computer Society Press, New York, 1994).

[3] L. K. Grover, "Quantum Mechanics Helps in Searching for a Needle in a Haystack", Phys. Rev. Lett. **79**, 325 (1997).

[4] M. A. Nielsen, I. L. Chuang, "Quantum Computation and quantum Information", (Cambridge University Press, Cambridge, 2000).

[5] A. Einstein, B. Podolsky, and N. Rosen, "Can Quantum-Mechanical Description of Physical Reality Be Considered Complete ?", Phys. Rev. **47**, 777 (1935).

[6] E. Schrödinger, "Discussion of probability relations between separated systems", Proceedings of the Cambridge Philosophical Society **31**, 555 (1935).

[7] A. Aspect, J. Dalibard and G. Roger, "Experimental Test of Bell's Inequalities Using Time- Varying Analyzers", Phys. Rev. Lett. **49**, 1804 (1982).

[8] C. Monroe, "Quantum information processing with atoms and photons", Nature **416**, 238 (2002).

[9] W. H. Zurek, "Decoherence, einselection, and the quantum origins of the classical", Rev. Mod. Phys. **75**, 715 (2003).

[10] P. R. Berman, "Cavity Quantum Electrodynamics", in Advances in Atomic, Molecular and Optical Physics, (Academic Press, New York, 1994).

[11] P. Goy, J. M. Raimond, M. Gross, and S. Haroche, "Observation of Cavity-Enhanced Single-Atom Spontaneous Emission", Phys. Rev. Lett. **50**, 1903 (1983).

[12] Y. Kaluzny, P. Goy, M. Gross, J. M. Raimond, and S. Haroche, "Observation of Self-Induced Rabi Oscillations in Two-Level Atoms Excited Inside a Resonant Cavity: The Ringing Regime of Superradiance", Phys. Rev. Lett. **51**, 1175 (1983).

[13] E. T. Jaynes and F. W. Cummings, "Comparison of quantum and semiclassical radiation theories with application to the beam maser", Proc. IEEE **51**, 89 (1963).

[14] P. M. Morse and H. Feshbach, "Methods of Theoretical Physics", (McGraw-Hill, New-York, 1953).

[15] E. Zauderer, "Partial differential equations of applied mathematics. Pure and applied mathematics", (Willey & Sons, New York, 1983).

[16] S. Nussmann et al., "Submicron Positioning of Single Atoms in a Microcavity", Phys. Rev. Lett. **95**, 173602 (2005).

[17] R. J. Thompson, G. Rempe, and H. J. Kimble, "Observation of normal-mode splitting for an atom in an optical cavity", Phys. Rev. Lett. **68**, 1132 (1992).

[18] K. M. Fortier et al., "Deterministic Loading of Individual Atoms to a High-Finesse Optical Cavity", Phys. Rev. Lett. **98**, 233601 (2007).

[19] P. Maunz, et al., "Normal-Mode Spectroscopy of a Single-Bound-AtomCavity System", Phys. Rev. Lett. **94**, 033002 (2005).

[20] M. Khudaverdyan et al., "Controlled insertion and retrieval of atoms coupled to a high-finesse optical resonator", New J. Phys. **10**, 073023 (2008).

[21] S.-B. Zheng and G.-C. Guo, "Efficient Scheme for Two-Atom Entanglement and Quantum Information Processing in Cavity QED", Phys. Rev. Lett. **85**, 2392 (2000).

[22] W. Dür, G. Vidal, and J. I. Cirac, "Three qubits can be entangled in two inequivalent ways", Phys. Rev. A **62**, 062314 (2000).

[23] A. Cabello, "Bells theorem with and without inequalities for the three-qubit Greenberger-Horne-Zeilinger and W states", Phys. Rev. A **65**, 032108 (2002).

[24] A. Sen(De), U. Sen, M. Wieśniak, D. Kaszlikowski, and M. Żukowski, "Multiqubit W states lead to stronger nonclassicality than Greenberger-Horne-Zeilinger states", Phys. Rev. A **68**, 062306 (2003).

[25] L. Ye, L. B. Yu, "Scheme for implementing quantum dense coding using tripartite entanglement in cavity QED", Phys. Lett. A **346**, 330 (2005).

[26] B. S. Shi and A. Tomita, "Teleportation of an unknown state by W state", Phys. Lett. A **296**, 161 (2002).

[27] C. Hai-Jing and S. He-Shan, "Quantum Key Distribution Using Four-Qubit W State", Commun. Theor. Phys. **46**, 65 (2006).

[28] M. Eibl et al., "Experimental Realization of a Three-Qubit Entangled W State", Phys. Rev. Lett. **92**, 077901 (2004).

[29] N. Kiesel et al., "Three-photon W-state", J. of Modern Optics **50**, 1131 (2003).

[30] G. Teklemariam et al., "Quantum erasers and probing classifications of entanglement via nuclear magnetic resonance", Phys. Rev. A **66**, 012309 (2002).

[31] F. Roos et al., "Control and Measurement of Three-Qubit Entangled States", Science **304**, 1478 (2004).

[32] L. You, X. X. Yi, and X. H. Su, "Quantum logic between atoms inside a high-Q optical cavity", Phys. Rev. A **67**, 032308 (2003).

[33] M. Weidinger, B. T. H. Varcoe, R. Heerlein, and H. Walther, "Trapping States in the Micromaser", Phys. Rev. Lett. **82**, 3795 (1999).

[34] J. M. Raimond, M. Brune, and S. Haroche, "Manipulating qunatum entanglement with atoms and photons in a cavity", Rev. Mod. Phys. **73**, 565 (2001).

[35] S. Haroche, J. M. Raimond, "Exploring the Quantum: Atoms, Cavities, and Photons", (Oxford Univeristy Press, 2006).

[36] D. M. Greenberger, M. Horne, and A. Zeilinger, in "Bells Theorem, Quantum Theory, and Conceptions of the Universe", edited by M. Kafatos, 73 (Kluwer, Dordrecht, 1989).

[37] M. Brune et al., "Quantum Rabi Oscillation: A Direct Test of Field Quantization in a Cavity", Phys. Rev. Lett. **76**, 1800 (1996).

[38] E. Hagley et al., "Generation of Einstein-Podolsky-Rosen pair of atoms", Phys. Rev. Lett. **79**, 1 (1997).

[39] S. Osnaghi et al., "Coherent Control of an Atomic Collision in a Cavity", Phys. Rev. Lett. **87**, 037902 (2001).

[40] A. Rauschenbeutel et al., "Coherent Operation of a Tunable Quantum Phase Gate in Cavity QED", Phys. Rev. Lett. **83**, 5166 (1999).

BIBLIOGRAPHY

[41] A. Rauschenbeutel et al., "Step-by-Step Engineered Multipartie Entanglement", Science **288**, 2024 (2000).

[42] A. Rauschenbeutel et al., "Controlled entanglement of two field modes in a cavity quantum electrodynamics experiment", Phys. Rev. A **64**, 050301(R) (2001).

[43] G. Nogues et al., "Absorption-free detection of a single photon", Nature **400**, 239 (1999).

[44] C. Guerlin et al., "Progressive field-state collapse and quantum non-demolition photon counting", Nature **448**, 889 (2007).

[45] S. Deleglise et al., "Reconstruction of non-classical cavity field states with snapshots of their decoherence ", Nature **455**, 510 (2008).

[46] M. Brune et al., "Process Tomography of Field Damping and Measurement of Fock State Lifetimes by Quantum Nondemolition Photon Counting in a Cavity", Phys. Rev. Lett. **101**, 240402 (2008).

[47] C. Schön, E. Solano, F. Verstraete, J. I. Cirac, and M. M. Wolf, "Sequential Generation of Entangled Multiqubit States", Phys. Rev. Lett. **95**, 110503 (2005); C. Schön, K. Hammerer, M. M. Wolf, J. I. Cirac, and E. Solano, "Sequential generation of matrix-product states in cavity QED", Phys. Rev. A **75**, 032311 (2007).

[48] H. J. Briegel and R. Raussendorf, "Persistent Entanglement in Arrays of Interacting Particles", Phys. Rev. Lett. **86**, 910 (2001).

[49] W. Dür and H.-J. Briegel, "Stability of Macroscopic Entanglement under Decoherence", Phys. Rev. Lett. **92**, 180403 (2004).

[50] V. Scarani, A. Acín, E. Schenck, and M. Aspelmeyer, "Nonlocality of cluster states of qubits", Phys. Rev. A **71**, 042325 (2005).

[51] X.-W. Wang, Y.-G. Shan, L.-X. Xia, and M.-W. Lu, "Dense coding and teleportation with one-dimensional cluster states", Phys. Lett. A **364**, 7 (2007).

[52] R. Raussendorf, D. E. Browne, and H. J. Briegel, "Measurement-based quantum computation on cluster states", Phys. Rev. A **68**, 022312 (2003).

[53] P. Walther et al., "Experimental one-way quantum computing", Nature **434**, 169 (2005).

[54] Y. Tokunaga, S. Kuwashiro, T. Yamamoto, M. Koashi and N. Imoto, "Generation of High-Fidelity Four-Photon Cluster State and Quantum-Domain Demonstration of One-Way Quantum Computing", Phys. Rev. Lett. **100**, 210501 (2008).

[55] X. B. Zou and W. Mathis, "Schemes for generating the cluster states in microwave cavity QED", Phys. Rev. A **72**, 013809 (2005).

[56] P. Dong, Z.-Y. Xue, M. Yang, and Z.-L. Cao, "Generation of cluster states", Phys. Rev. A **73**, 033818 (2006).

[57] J.-Q. Li, G. Chen, and J.-Q. Liang, "One-step generation of cluster states in microwave cavity QED", Phys. Rev. A **77**, 014304 (2008).

[58] R. ul-Islam, "Engineering Entanglement in Cavity Quantum Electrodynamical Systems", (Lambert Academic Publishing, 2010).

[59] X.-W. Wang and G.-J. Yang, "Schemes for preparing atomic qubit cluster states in cavity QED", Opt. Commun. **281**, 5282 (2008).

[60] D. Gonta, S. Fritzsche, and T. Radtke, "Generation of four-partite Greenberger-Horne-Zeilinger and W states by using a high-finesse bimodal cavity", Phys. Rev. A **77**, 062312 (2008).

[61] D. Gonta, S. Fritzsche, and T. Radtke, "Generation of two-dimensional cluster states by using high-finesse bimodal cavities", Phys. Rev. A **79**, 062319 (2009).

[62] D. Gonta and S. Fritzsche, "Control of entanglement and two-qubit quantum gates with atoms crossing a detuned optical cavity", J. Phys. B **42**, 145508 (2009).

[63] D. Gonta and S. Fritzsche, "Multipartite W states for chains of atoms conveyed through an optical cavity", Phys. Rev. A **81**, 022326 (2010).

[64] R. H. Dicke, "Coherence in Spontaneous Radiation Processes", Phys. Rev. **93**, 99 (1954).

[65] N. Schuch and J. Siewert, "Natural two-qubit gate for quantum computation using the XY interaction", Phys. Rev. A **67**, 032301 (2003).

[66] A. Yariv, "Quantum Electronics", (Wiley, New. York, 1989).

[67] R. D. Günther, "Modern Optics", (Wiley, New. York, 1990).

[68] A. Zeilinger, "Experiment and the foundations of quantum physics", Rev. Mod. Phys. **71**, S288 (1999).

BIBLIOGRAPHY

[69] Q. A. Turchette, C. J. Hood, W. Lange, H. Mabuchi, H. J. Kimble, "Measurement of Conditional Phase Shifts for Quantum Logic", Phys. Rev. Lett. **75**, 4710 (1995).

[70] L. M. Duan and H. J. Kimble, "Scalable Photonic Quantum Computation through Cavity-Assisted Interactions", Phys. Rev. Lett. **92**, 127902 (2004).

[71] S. Kuhr et al., "Ultrahigh finesse Fabry-Perot superconducting resonator", Appl. Phys. Lett. **90**, 164101 (2007).

[72] A. Biswas, G. S. Agarwal, "Preparation of W, GHZ, and two-qutrit states using bimodal cavities", J. of Modern Optics **51**, 1627 (2004).

[73] M. Ikram, F. Saif, "Engineering entanglement between two cavity modes", Phys. Rev. A **66**, 014304 (2002).

[74] C. Wildfeuer and D. H. Schiller, "Generation of entangled N-photon states in a two-mode Jaynes-Cummings model", Phys. Rev. A **67**, 053801 (2003).

[75] M. S. Zubairy, M. Kim, and M. .O. Scully, "Cavity-QED-based quantum phase gate", Phys. Rev. A **68**, 033820 (2003).

[76] I. P. de Queiros, S. Souza, W. B. Cardoso, and N. G. de Almeida, "Accuracy of a teleported trapped field state inside a single bimodal cavity", Phys. Rev. A **76**, 034101 (2007).

[77] D. Gonta and S. Fritzsche, "On the role of the atom-cavity detuning in bimodal cavity experiments", J. of Phys. B **41**, 090101 (2008).

[78] A. R. Bosco de Magalhaes and M. C. Nemes, "Searching for decoherence-free subspaces in cavity quantum electrodynamics", Phys. Rev. A **70**, 053825 (2004).

[79] M. Khudaverdyan et al., "Quantum Jumps and Spin Dynamics of Interacting Atoms in a Strongly Coupled Atom-Cavity System", Phys. Rev. Lett. **103**, 123006 (2009).

[80] S. Kuhr et al., "Deterministic Delivery of a Single Atom", Science **293**, 278 (2001).

[81] H. Carmichael, "An Open System Approach to Quantum Optics", (Springer Verlag Berlin Heidelberg, 1993).

[82] P. Meystre and M. Sargent, "Elements of Quantum Optics", (Springer Verlag Berlin Heidelberg, 2007).

[83] R. G. Hulet and D. Kleppner, "Rydberg Atoms in Circular States", Phys. Rev. Lett. **51**, 1430 (1983).

[84] T. F. Gallagher, "Rydberg atoms", (Cambridge University Press, 1994).

[85] J. I. Kim et al., "Classical Behavior with Small Quantum Numbers: The Physics of Ramsey Interferometry of Rydberg Atoms", Phys. Rev. Lett. **82**, 4737 (1999).

[86] G.-P. Guo, C.-F. Li, J. Li, and G.-C. Guo, "Scheme for the preparation of multi-particle entanglement in cavity QED", Phys. Rev. A **65**, 042102 (2002).

[87] M. Amniat-Talab, S. Guérin, N. Sangouard, and H. R. Jauslin, "Atom-photon, atom-atom, and photon-photon entanglement preparation by fractional adiabatic passage", Phys. Rev. A **71**, 023805 (2005).

[88] C. Marr, A. Beige, and G. Rempe, "Entangled-state preparation via dissipation-assisted adiabatic passages", Phys. Rev. A **68**, 033817 (2003).

[89] Y. Li, C. Bruder, and C. P. Sun, "Time-dependent Fro"hlich transformation approach for two-atom entanglement generated by successive passage through a cavity", Phys. Rev. A **75**, 032302 (2007).

[90] G. R. Guthöhrlein, M. Keller, K. Hayasaka, W. Lange, and H. Walther, "A single ion as a nanoscopic probe of an optical field", Nature **414**, 49 (2001); M. Keller, B. Lange, K. Hayasaka, W. Lange, and H. Walther, "Continuous generation of single photons with controlled waveform in an ion-trap cavity system", Nature **431**, 1075 (2004).

[91] A. B. Mundt et al., "Coupling a Single Atomic Quantum Bit to a High Finesse Optical Cavity", Phys. Rev. Lett. **89**, 103001 (2002); A. Kreuter et al., "Spontaneous Emission Lifetime of a Single Trapped Ca^+ Ion in a High Finesse Cavity", Phys. Rev. Lett. **92**, 203002 (2004).

[92] R. Drever and J. Hall, "Laser phase and frequency stabilization using an optical resonator", Appl. Phys. B **31**, 97 (1983).

[93] H. J. Metcalf and P. van der Straten, "Laser Cooling and Trapping", (Springer-Verlag New York, 1999).

[94] C. J. Foot, "Atomic Physics", (Oxford University Press, 2005).

[95] E. Brion, L. H. Pedersen and K. Mølmer, "Adiabatic elimination in a lambda system", J. Phys. A: Math. Theor. **40**, 1033 (2007).

[96] D. A. Steck, "Cesium D Line Data", http://steck.us/alkalidata/.

BIBLIOGRAPHY

[97] J. P. Gordon and A. Ashkin, "Motion of atoms in a radiation trap", Phys. Rev. A **21**, 1606 (1980).

[98] L. Förster et al., "Number-triggered loading and collisional redistribution of neutral atoms in a standing wave dipole trap", New J. Phys. **8**, 259 (2006).

[99] Y. Miroshnychenko et al., "An atom-sorting machine", Nature **442**, 151 (2006).

[100] D. Schrader et al., "An optical conveyor belt for single neutral atoms", Appl. Phys. B **73**, 819 (2001).

[101] S. Kuhr et al., "Coherence properties and quantum state transportation in an optical conveyor belt", Phys. Rev. Lett. **91**, 213002 (2003).

[102] D. Schrader et al., "Neutral atom quantum register", Phys. Rev. Lett. **93**, 150501 (2004).

[103] R. Grimm, M. Weidemüller, and Y. B. Ovchinnikov, "Optical dipole traps for neutral atoms", Adv. At. Mol. Opt. Phys. **42**, 95 (2000).

[104] B. Weber et al., "Photon-Photon Entanglement with a Single Trapped Atom", Phys. Rev. Lett. **102**, 030501 (2009).

[105] S. Nußmann et al., "Vacuum-stimulated cooling of single atoms in three dimensions", Nat. Phys. **1**, 122 (2005).

[106] T. Wilk, S. C. Webster, H. P. Specht, G. Rempe, A. Kuhn, "Polarization-controlled single photons", Phys. Rev. Lett. **98**, 063601 (2007).

[107] H. P. Specht et al., "Shaping the Phase of a Single Photon", Nat. Phot. **3**, 469 (2009).

Die VDM Verlagsservicegesellschaft sucht für wissenschaftliche Verlage abgeschlossene und herausragende

Dissertationen, Habilitationen, Diplomarbeiten, Master Theses, Magisterarbeiten usw.

für die kostenlose Publikation als Fachbuch.

Sie verfügen über eine Arbeit, die hohen inhaltlichen und formalen Ansprüchen genügt, und haben Interesse an einer honorarvergüteten Publikation?

Dann senden Sie bitte erste Informationen über sich und Ihre Arbeit per Email an *info@vdm-vsg.de*.

Sie erhalten kurzfristig unser Feedback!

VDM Verlagsservicegesellschaft mbH
Dudweiler Landstr. 99
D - 66123 Saarbrücken
Telefon +49 681 3720 174
Fax +49 681 3720 1749
www.vdm-vsg.de

Die VDM Verlagsservicegesellschaft mbH vertritt

Printed by Books on Demand GmbH, Norderstedt / Germany